Routledge Introductions to Environment

Environment and Politics

Timothy Doyle and
Doug McEachern

London and New York

First published 1998
by Routledge
11 New Fetter Lane, London EC4P 4EE

Simultaneously published in the USA and Canada
by Routledge
29 West 35th Street, New York, NY 10001

Typeset in Times by Keystroke, Jacaranda Lodge, Wolverhampton
Printed and bound in Great Britain by Biddles Ltd, Guildford and King's Lynn

British Library Cataloguing in Publication Data
A catalogue record for this book is available from the British Library

Library of Congress Cataloguing in Publication Data
Doyle, Timothy, 1960–
    Environment and politics / Timothy Doyle and Douglas McEachern.
      p.   cm. — (Routledge introductions to environment)
    Includes bibliographical references and index.
      1. Environmental policy. 2. Green movement—Political aspects.
    I. McEachern, Doug. II. Title. III. Series.
GE170. D69   1998
363.7—dc21   97-17918

ISBN 0–415–14775–1 (hbk)
      0–415–14776–x

**Books are to be returned on or before
the last date below.**

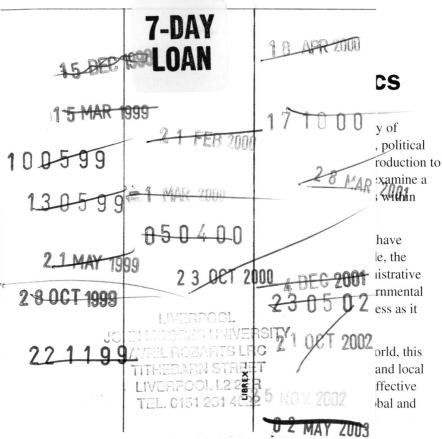

CS

y of
, political
roduction to
examine a
within

have
le, the
iistrative
rnmental
ess as it

orld, this
and local
ffective
bal and

Timothy Doyle is a Senior Lecturer at the Mawson Graduate Centre for
Environmental Studies, University of Adelaide. **Doug McEachern** is
Professor of Politics at the University of Adelaide.

Routledge Introductions to Environment Series
Published and Forthcoming Titles

Titles under Series Editors:
Rita Gardner and Antoinette Mannion

Environmental Science texts

Environmental Biology
Environmental Chemistry and Physics
Environmental Geology
Environmental Engineering
Environmental Archaeology
Atmospheric Systems
Hydrological Systems
Oceanic Systems
Coastal Systems
Fluvial Systems
Soil Systems
Glacial Systems
Ecosystems
Landscape Systems

Titles under Series Editors:
David Pepper and Phil O'Keefe

Environment and Society texts

Environment and Economics
Environment and Politics
Environment and Law
Environment and Philosophy
Environment and Planning
Environment and Social Theory
Environment and Political Theory
Business and Environment

Key Environmental Topics texts

Biodiversity and Conservation
Environmental Hazards
Natural Environmental Change
Environmental Monitoring
Climatic Change
Land Use and Abuse
Water Resources
Pollution
Waste and the Environment
Energy Resources
Agriculture
Wetland Environments

Energy, Society and Environment
Environmental Sustainability
Gender and Environment
Environment and Society
Tourism and Environment
Environmental Management
Environmental Values
Representations of the Environment
Environment and Health
Environmental Movements
History of Environmental Ideas
Environment and Technology
Environment and the City
Case Studies for Environmental Studies

Contents

Series editors' preface
Environment and Society titles

The 1970s and early 1980s constituted a period of intense academic and popular interest in processes of environmental degradation: global, regional and local. However, it soon became increasingly clear that reversing such degradation would not be a purely technical and managerial matter. All the technical knowledge in the world does not necessarily lead societies to change environmentally damaging behaviour. Hence a critical understanding of socio-economic, political and cultural processes and structures has become, it is acknowledged, of central importance in approaching environmental problems. Over the past two decades in particular there has been a mushrooming of research and scholarship on the relationships between social sciences and humanities on the one hand and processes of environmental change on the other. This has lately been reflected in a proliferation of associated courses at undergraduate level.

At the same time, changes in higher education in Europe, which match earlier changes in America, Australasia and elsewhere, mean that an increasing number of such courses are being taught and studied within a framework offering maximum flexibility in the typical undergraduate programme: 'modular' courses or their equivalent.

The volumes in this series will mirror these changes. They will provide short, topic-centred texts on environmentally relevant areas, mainly within social sciences and humanities. They will reflect the fact that students will approach their subject matter from a great variety of different disciplinary backgrounds; not just within social sciences and humanities, but from physical and natural sciences too. And those students may not be familiar with the background to the topic, they may or may not be going on to develop their interest in it, and they cannot automatically be thought of as being at 'first-year level', or second- or third-year: they might need to study the topic in any year of their course.

The authors and editors of this series are mainly established teachers in higher education. Finding that more traditional integrated environmental studies or specialised academic texts do not meet their requirements, they have increasingly met the new challenges caused by structural changes in education by writing their own course materials for their own students. These volumes represent, in modified form which all students can now share, the fruits of their labours.

To achieve the right mix of flexibility, depth and breadth, the volumes, like most modular courses themselves, are designed carefully to create maximum accessibility to readers from a variety of backgrounds. Each leads into its topic by giving adequate introduction, and each 'leads out' by pointing towards complexities and areas for further development and study. Indeed, much of the integrity and distinctiveness of the Environment and Society titles in the series will come through adopting a characteristic, though not inflexible, structure to the volumes. Each introduces the student to the real-world context of the topic, and the basic concepts and controversies in social science/humanities which are most relevant. The core of each volume explores the main issues. Data, case studies, overview diagrams, summary charts and self-check questions and exercises are some of the pedagogic devices that will be found. The last part of each volume will normally show how the themes and issues presented may become more complicated, presenting cognate issues and concepts needing to be explored to gain deeper understanding. Annotated reading lists are important here.

We hope that these concise volumes will provide sufficient depth to maintain the interest of students with relevant backgrounds, and also sketch basic concepts and map out the ground in a stimulating way for students to whom the whole area is new.

The Environment and Society titles in the series complement the Environmental Science titles which deal with natural science-based topics. Together this comprehensive range of volumes which make up the Routledge Introductions to Environment Series will provide modular and other students with an unparalleled range of perspectives on environmental issues, cross referencing where appropriate.

The main target readership is introductory level undergraduate students predominantly taking programmes of environmental modules. But we hope that the whole audience will be much wider, perhaps including second- and third-year undergraduates from many disciplines within the social sciences, science/technology and humanities, who might be taking occasional environmental courses. We also hope that sixth-form teachers and the wider public will use these volumes when they feel the need to obtain quick introductory coverage of the subject we present.

David Pepper and Phil O'Keefe
1997

Series International Advisory Board

Australasia: Dr P. Curson and Dr P. Mitchell, Macquarie University

North America: Professor L. Lewis, Clark University; Professor L. Rubinoff, Trent University

Europe: Professor P. Glasbergen, University of Utrecht; Professor van Dam-Mieras, Open University, The Netherlands

Plates

Figures

Tables

Boxes

Authors' preface

Environmental politics has many different faces. It is a world of contrasts and comparisons. Consider, for example, the contrast in a single day in Cape Town, South Africa. In the morning, early, we rose and drove out towards the east coast to visit Rondevlei Nature Reserve. This was set up in 1952 to conserve a wetland and its birdlife. It now lies behind a large mesh fence separating it from the 'coloured' area of Grassy Park and a black settlement. It is a good example of old-style wildlife conservation and, in its way, symbolic of the apartheid era. The focus of Rondevlei is well stated in its publicity brochure.

> Although Rondevlei's original purpose was to conserve the rich birdlife associated with this important wetland, the emphasis has shifted to the conservation of the indigenous flora, fauna and natural features of the area, while still offering people a place where they can enjoy nature.
>
> Rondevlei has also become an important environmental education centre, being within easy reach of the many schools in and around the Peninsula.

It is a good bird sanctuary, preserving the habitats against urban encroachment and rescuing habitat from invasive exotic plants. There is even a population of hippopotamuses, although these are not often seen by the casual visitor. The emphasis is on wildlife and nature conservation and the people of the area come as visitors, outsiders 'to enjoy nature'.

In the afternoon, we travelled out to Langa, one of the oldest black townships, and visited the Tsoga Environment Centre. Langa was shaped and blighted by the power of the state and the apartheid imagination but possesses a civic spirit built on the pride of the African population in resistance to that malignant vision. Here in some recycled shipping containers is a thriving civic organisation dedicated to the greening of Langa. This group has developed a permaculture garden, and sponsored the development of local vegetable gardens and the planting of flowers on the verges and front gardens in the centre of the township. Tsoga confronts environmental degradation where it hits the everyday lives of the citizens of the new South Africa. Not everyone in Langa is enthusiastic about the goals of the group, as good housing and employment are the most pressing needs of the local community, but with great skill, commitment and dedication Tsoga is slowly making a difference, planting the appropriate trees to lessen need for the *sangomas* to collect healing plant materials from the wild and creating parks where before there were dusty open spaces.

Rondevlei and Tsoga show two different ways of responding to threats to the environment and human wellbeing. In this book we have sought to explore these and many other ways of doing environmental politics and to consider the different ways in which power is used to either damage or protect the environment. The contrasts are great, but it is by considering the contrasts and similarities across the globe that it is possible to come to some sense of what is going on and what is at stake in the political struggle for our environmental future.

Tim Doyle
Doug McEachern
April 1997

Acknowledgements

Tim would like to thank the Department of Government, Clark University, Mass., USA; the Environmental Studies Program, Montana, USA; Mawson Graduate Centre for Environmental Studies, University of Adelaide; Ton Bührs; Brian Cook; Tom Roy; FOE Fitzroy; Chain Reaction.

Tim would like to dedicate this volume to Fiona.

Doug would like to thank the Department of Politics, University of Adelaide; the Department of Political Science, University of Cape Town; the Research School of Social Science, Australian National University; and the Tsoga Environment Centre, Langa, South Africa.

We would both like to thank the students of the various courses we have taught on environmental politics at the above universities for their demand for better-quality information and argument on these topics, and our colleagues who encouraged us in this project. We would both like to thank Mike Howes. We would like to thank Sarah Lloyd of Routledge for her assistance, which went beyond that normally expected of an editor. We would also like to express our appreciation for the efforts of all those involved in environmental struggles, continually pushing forward their understanding of how environmental damage is caused and cured. It was their argument and efforts which made a text such as this both necessary and possible.

Introduction

Although environmental issues and forms of environmental concern have a very long history, awareness of the environmental consequences of economic development was given an increasingly political character from the 1950s onwards (Young 1990). Individuals produced provocative studies warning of particular threats to the environment, as with Rachel Carson's well-known criticism of the increased use of DDT as a pesticide. Groups formed to press for solutions to particular or local problems or sought to get the political system to respond. Think tanks, such as the Club of Rome, published accounts dramatising the potential depletion of the Earth's resources. International agencies, including the United Nations Environment Programme, began holding international conferences and promoting detailed studies of issues as part of an effort to get more co-ordinated and effective responses to increasingly global environmental problems. Later, protest movements, linking up with late 1960s student radicalism and with various anti-war mobilisations, took to the streets and forests in efforts to get a political response. In some places the mainstream political parties began to respond; in others, environmental concern was mocked and marginalised. In West Germany, a history of radical protest in the midst of obvious environmental problems and the nuclear threat of an active phase in the Cold War produced the formation of a radical Green Party (Die Grünen), which from 1980 to 1982 had enough electoral support to be represented in various state parliaments and in 1983 to enter the Federal Parliament in Bonn (Young 1990: 171). The image of the West German Greens strolling into parliament in casual clothes, carrying potted plants and announcing that they were there to represent the politics of life was a sufficiently dramatic symbol to suggest that new forms of environmental politics were starting to challenge routine assumptions about the workings of the 'normal' political process.

The purpose of this book is to provide an outline of the concepts that can be used to analyse and assess the character and consequences of environmental politics, in all its varied forms. In the course of presenting these concepts and the arguments about them, a broad outline of the development of these forms of environmental politics has also been provided. Most books on the environment begin with a litany of environmental problems and issues, but the very concept of the environment has been used in so many and in such contested ways that its meaning has become quite problematic. What counts as an environmental problem depends on varying ways of judging the ecological consequences of any particular act or development. It is important to understand that the term 'environment' is often constructed differently in different cultures and is used in different ways by different people. Not only is the word defined differently, but alternative clusters of issues are identified and a whole range of different kinds of politics can be generated on this basis. For example, environmental politics has been used to challenge the status quo in many societies. In North America, Australia and parts of Scandinavia the environmental agenda has often been dominated by attempts to protect wilderness areas from the intrusion and excesses of human development. Here environmental conflict has challenged the dominant goals of advanced capitalism and industrialism, such as unlimited growth and the rights of private property. It would be wrong to view environmental politics simply as a challenge to capitalist orthodoxy. In Eastern Europe, environmental politics developed as a rejection of state socialism's promotion of rapid and ecologically damaging forms of industrial and agricultural development. As part of the East European 'velvet revolutions' of the late 1980s, environmentalists championed pluralist democracy and 'free-market' economic solutions derived from the tenets of capitalism, in a bid to overthrow decades of rigid, bureaucratic and authoritarian rule.

Some Asian and African countries have used environmental debates to challenge a different and global status quo, where a few affluent and powerful countries have access to, and consume, disproportionately greater amounts of the Earth's limited resources. The governments of these countries are less concerned with the rights of 'other nature' and are more concerned with promoting economic development to raise the standard of living of their populations. In these countries, environmental politics often focuses on issues of human survival such as the adequate provision of housing, food and employment, as well as safe work and healthy living conditions.

As far as these governments are concerned, the more powerful nations are busy promoting their own definitions of environmental ills, and promoting their own plans to solve these problems. 'Global ecology' is an excellent example of the environmental agenda of the 'developed world', since the global ecological issues on top of the North's environmental agenda are often not the same as those issues espoused by governments and peoples of the South.[1] For example, population control, species extinction, global climate change and deforestation are high-priority problems as defined in Northern elite and scientific terms. In the South, the emphasis is on solving environmental problems that have an impact on basic levels of standard of living and quality of life.

Governments of some industrialising countries oppose moves from the USA and Europe to impose global environmental objectives on them. Hence, in the international debates over greenhouse problems, Malaysia has set itself firmly against any moves that would make it harder for it to industrialise and export its goods. It should be noted that within these industrialising/developing countries environmental issues are still raised by groups of environmental activists. For example, in Nigeria, the struggle of the Ogoni people against the military regime has a strong environmental theme. Here, poor people protest against the damage being done to their land by the polluting activities of Shell Nigeria and the links between Shell Nigeria and the oppressive military regime. In Indonesia, environmental groups oppose the logging of rain forests; in India, environmentalists oppose excessive logging of forests, the polluting consequences of industrialisation, and the environmental damage associated with population growth and urbanisation, and they are concerned about the consequence of global climate change. Further, in many parts of the world environmental politics has been used to contest inequalities and differences in power based on gender, class, race and species.

Although the concept of 'the environment' is invoked to support the struggles of the less powerful it can also be used by those either in or with power. It must be understood that environmental symbols are also used by the powerful to push their own interests. For example, some powerful business interests have also managed to enter the environmental 'corner', sometimes arguing that a healthy environment is one and the same as a healthy, expanding economy, with a marketplace unfettered by governmental controls and national boundaries. In fact, with 'sustainable development' a 'good' environment is actually 'good for business'. In this way, much environmental politics becomes oriented around efficiency and effectiveness criteria, consolidating the power of 'business as usual'.

So environmental politics is not just about 'goodies' versus 'baddies'. This symbol *environment* has such power that numerous cultures, and the powerful and powerless within them, invoke its name for disparate purposes. The definition of environment used may determine whether or not one is willing to accept the existence of an *environmental crisis*. For example, some environmentalists would argue that to speak of a 'crisis' is an unproductive, meaningless exaggeration. Environmental problems are not systematically linked and it is appropriate to make incremental adjustments to normal operating procedures as a response to evidence of particular pieces of ecological damage. In this way, 'sustainable development' constructs all environment 'problems' as efficiency issues, which have to be *managed* more effectively. These management issues include being more technologically, economically, organisationally, educationally and politically efficient. By constructing 'end-of-pipe' technologies to prevent waste products entering the environment as pollution or by redesigning the production process, these efficiency problems can be largely resolved. On the east coast of the United States, the 'environment' is almost entirely constructed in terms of 'air', 'land' and 'water' issues, and understood in these terms. Incrementally, through adequate recycling technology, among other things, all environmental problems can, in time and with effort, be solved. Greg Easterbrook (1995) is one who argues that all the hard decisions have already been made and from now on we just need to be better managers. He argues that:

- In the Western world pollution will end in our lifetimes;
- First world industrial countries are cleaner than developing countries;
- Most feared environmental catastrophes, like global warming, will be avoided; and
- There is no conflict between the artificial and the natural.

Easterbrook dismisses those critics who suggest that there are serious environmental issues that cannot be addressed through 'environmental best practice' management as 'Cassandras'. Often these environmental critics do use the notion of environmental crisis and portray the Easterbrooks of the world as 'Pollyannas'. The whole debate between

Pollyannas (overly 'positive') and Cassandras (overly 'negative') hinges on what depiction/construction of 'the environment' is being projected. The sustainable development 'Pollyanna' sees 'the environment' as external to human beings, an instrumental resource to be used and managed for human purposes but with a capacity to have impacts upon people and vice versa. It does not see humanity as intrinsically part of the environment or nature. Where humans are seen as 'intrinsic' to nature (by the Cassandras), environmental problems appear somewhat larger and more crisis-like. Issues include poverty, homelessness, disease, and diminished diversity in terms of languages, lifestyles, ideas and political forms, as well as the loss of diversity in 'other nature'; the increasing gap between the haves and the have nots; the poisoning of the Earth through chemicals and other toxins; the extreme degradation inflicted upon the Earth by all versions of advanced industrialism, and the lingering threat of nuclear disaster.

There is a multiplicity of green political responses to the myriad of environmental problems around the globe. With the need for environmental problem solving to be both global and region-specific, some patterns emerge pointing to international solutions. Global solutions cannot simply emerge from the administrative documents generated by international diplomatic environmental forums. Green political solutions are partly to be found in the experiences of different political cultures. Equally important is the knowledge and respect shown by international environmental policy-makers for these national and local traditions.

Let us now turn to a brief outline of the structure of the book and its order of exposition.

Chapter 1 outlines what is meant by the central concepts of politics and environmental studies. The different possible definitions of each concept vastly affect relations between the two, shaping what needs to be understood and the terms in which the analysis will be conducted. Power is basic to both the practice and the assessment of politics. This chapter presents a series of different definitions of power, the models of society on which they are based and the different methods they suggest for the study of environmental politics, conflict and policy making.

There have been three broad types of political response to environmental concern: to reject and resist it; to accommodate it, and propose reform; or to embrace it and demand revolutionary or radical change. Each of these responses is associated with an ideology or 'world view' made up of ideas and value judgements. Sometimes these ideologies are coherent and cleanly structured; on other occasions they are a jumble of ambiguous and contradictory values and beliefs.[2] Chapter 2 provides a summary of the key versions of these political responses and their world views.

This book is conceived as an introduction to the comparative study of environmental politics. It is assumed that what happens in environmental politics varies between countries on the basis of the different political systems they possess, the different kinds of environmental problems experienced and the severity of those problems, the differing strength and capacity of governments to make and enforce environmental decisions, and differences in culture. The broad concepts needed to study environmental politics in any of these countries may be the same but their application will provide quite varied interpretations of why things have happened in the way that they have as well as contrasting evaluations of the significance of different outcomes. Both elements of similarity and differences will be stressed.

Apart from providing a comparative look at different nations' experiences with environmental politics, this book is also comparative in its investigation of different types of policy process within these countries. Policy is not a document but a political process. Numerous processes can be used to create, sustain or thwart environmental change. Often, political structures determine what types of environmental goals can be pursued. In addition, depending on what the environmental goals are, radical or reformist, certain political structures are more useful than others. For example, environmental politics is played quite differently in informal networks within social movements than within more formalised non-governmental organisations. It is played differently in political parties than in bureaucracies and administrative systems. It is played differently in business corporations than in elected governments. Each kind of organisation produces different approaches to environmental politics and each form interacts with other structures of politics and policy making. Chapters 3 to 8 present accounts of the varying ways in which different political processes have responded to the 'greening of society'.

Chapter 3 presents different ways of analysing the general character of social movements, with particular attention paid to the specific attributes of environmental social movements. What are social movements? Why do they emerge in particular countries at particular times? How are they structured? What is their potential for promoting or resisting environmental change? What kinds of politics are associated with these new social movements? How does the form of politics vary in different parts of the world? This chapter emphasises the importance of informal politics, which often cuts across national, cultural and institutional boundaries.

Non-governmental organisations (NGOs) are the subject of Chapter 4. NGOs are the most visible players in environmental politics around the globe. They are involved in many different spheres of politics, from the local community level through to the politics of the nation-state and inter- or trans-national politics. They exist in both the predominantly non-institutional domain of social movement politics, and the institutionalised milieu of political parties, administrative systems, governments and beyond.

Chapter 5 concentrates on green politics and the formation of green political parties. The focus here is primarily on events in the liberal democracies of Western Europe, North America and Australasia. Consideration is given to the various factors that shape the development of green parties and allow them a degree of electoral success. The most significant example is, of course, the case of the West German Greens, but there are other examples of green party formation and political success inside the standard processes of representative politics. This chapter also explores those factors that inhibit the development and success of these parties.

Many works that analyse politics pay little attention to the role played by organised business. It is sometimes assumed that politics takes place in a distinct public sphere and that business is confined to the economy, but on questions of the environment the business of the marketplace has a strong political dimension. Particular firms and various business associations can be active in defending themselves against various environmental campaigns and the threat of environmental regulation. At times, business has been the key player in trying to define an acceptable approach to the making of policy on environmental questions. Much of the content for the concept of sustainable development has come from the efforts of business and the research it has sponsored. This is even more true of the development and dissemination of 'free market' arguments about how to respond to environmental problems. Chapter 6 explores both the various power resources deployed by business and the strategies developed by different parts of business to protect themselves and the right to pursue 'business as usual'.

Chapter 7 discusses institutional politics and policy making. When an environmental issue is put on the political agenda and government chooses to respond, that response will be enacted, regulated and monitored by the administrative arm of the state. It is not just that the bureaucracy does what the political process tells it, since important information and policy advice is generated by the bureaucracy itself. The very character of the policy response may well be determined by what the bureaucracy does. In recent years, there has been a major debate about the appropriateness and efficiency of direct governmental regulation of environmental problems as well as strong advocacy of the use of prices and markets to produce better environmental outcomes. These important questions are also considered in Chapter 7.

Environmental politics cannot be treated only as it happens within single, separate nation-states. Issues are shared between nations, hence the scope for a comparative evaluation. More than that, some environmental issues cross national boundaries in the sense that action in one country produces environmental harm in another. Others are more global in character and require co-operative, negotiated and diplomatic efforts to generate even plausible responses. The final chapter explores what is needed to understand the success or failure of attempts to develop effective international regimes to deal with these complex global and cross-border environmental issues. It also considers a number of case studies to show what has been done in various major international forums, such as the Rio Earth Summit of 1992, and in the attempts to develop treaties in response to global issues, such as climate change and ozone depletion.

The book concludes with a survey of the broad analytical concepts required for a comparative understanding of environmental politics as well as a restatement of the importance of the comparative perspective.

Further Reading

Easterbrook, G. (1995) *A Moment on the Earth: The Coming Age of Environmental Optimism*, Viking Penguin, New York.

Seliger, M. (1976) *Ideology and Politics*, Allen & Unwin, London.

Young, J. (1990) *Post Environmentalism*, Belhaven Press, London.

1 Politics and environmental studies

- Central concepts of politics
- Political regimes
- Dimensions of power
- Models of policy processes

Introduction

Both environmental studies and the discipline of politics can be conceived of, and practised, in quite different ways. For example, environmental studies can be treated as a sub-discipline of natural science and bring to bear that whole approach to understanding particular problems and to evaluating alternative solutions. That is not the only approach. It is equally possible to understand environmental studies as essentially an interdisciplinary enterprise bringing together knowledge and expertise from a whole range of scientific and social science disciplines. The study of politics is capable of equally varied conceptions, ranging from its basic focus through to questions of method and means of evaluation. This chapter starts with a consideration of the delineation of the two fields of environmental studies and politics to establish what is at the core for the comparative study of environmental politics. On that basis some of the key, even if contested, concepts of politics are outlined along with sketches of alternative methodologies and assumptions about the workings of the different political processes found across the world today.

Defining politics and environmental studies

What is environmental studies?

Environmental studies can be approached in a number of significantly different ways. It is possible to treat the field as being coterminous with that of 'ecology', the study of the particular ways in which living matter combines in changing and stable patterns. Here ecology is a branch of science to be approached with the same methodological assumptions as other areas of scientific inquiry. Such an approach could generate studies of, for example, the ecology of a particular region (the redwood forests of North America, the moorlands of England, the Great Barrier Reef in Australia or the Karoo in South Africa). Even in such studies of nature and natural systems there would be some recognition of the part played by humans in this ecology. For example, human actions may destabilise a particular ecological balance and impose new, managed stability/instability, and this could be noted with the same kind of scientific detachment as for the study of other disturbances.

Recognising the human element indicates one of the problems for defining environmental studies only in ecological terms. How are the actions of humans and human societies to be interpreted? It is certainly possible to invent a 'human ecology' and treat this as a sub-set of other parts of scientific study: it could form part of zoology, biology or human geography. It is certainly possible to gain some understanding of the character of environmental problems by considering population dynamics and the causes and consequences of 'overpopulation', but how good or useful is this interpretation? There is more to understanding the impact and dynamics of human social organisation than can be generated from this particular perspective. Here it is possible to extend the conception of environmental studies and to enhance the interdisciplinary element. Ecology itself involves forms of interdisciplinary study. To understand the interaction of life forms means crossing, at the very least, the boundaries between botany, zoology and soil science. But this interdisciplinary focus needs to be enhanced to include, at the least, the social sciences if we want to understand and assess the environmental consequences of humans as socially organised.

From this perspective, environmental studies needs to combine understandings from both the sciences and social sciences and it needs to be interdisciplinary in this sense. Knowledge derived from ecology and a whole range of individual disciplines needs to be combined with

Plate 1 *Banner at protest against visit by nuclear powered/capable US ship, Williamstown, Victoria. Courtesy of Friends of the Earth*

knowledge derived from the social sciences. For example, human impacts on their environments depend on a whole range of factors: cultural attitudes to nature; modes of social and economic organisation; and the kinds of political process that can either protect or harm the environment.

Depending on how environmental studies is defined, the form of political knowledge required will be quite varied.

What is politics?

What is understood as 'politics' and 'political' varies widely. Often politics has been defined in a particularly narrow way and the word is used to refer to processes of government; decision making and administration; elections; the machinations of political parties; and the efforts of groups to influence these political processes. This limited, 'government-centred' view of politics, according to Crick, emerged in advanced, complex, usually European, societies:

> The establishment of political order is not just any order at all; it marks the birth, or the recognition, of freedom. For politics represents at least

some tolerance of different truths, some recognition that government is
possible, indeed best conducted, amid the open canvassing of rival
interests. Politics are the public actions of free men.

(1964: 18)

This view of politics, which sees government as a public instrument of
freedom, is associated with a 'Western' tradition, reaching back to the
ancient Greeks and illuminated in the writings of Aristotle and Plato.
Here politics is seen in a positive light, as part of the way in which
citizens are fulfilled and the highest goals of a community are
achieved. Involvement in public debate over common problems and
their solutions was seen as part of civic duty: to be involved in politics
was a high cultural and social ambition. In essence this is a conception
of politics as some variant of democratic politics. Today, democracy
and democratic politics are greatly valued but the practice of politics
has separated out the role of citizen from that of the politician.
Politicians are professionals elected to act on behalf of voters in
remote parliaments and governmental processes. It is largely their job
to reconcile conflicting and convergent societal interests through
compromise and negotiation in a public sphere. In this view, politics
exists 'out there'. Most citizens are removed from its daily reach.
Politics is regarded as that peculiar set of relationships forged in the
parliaments and congresses of national and state capitals. It is important to note that Crick's limited view of politics is shared by most
people. Given this alienation, politics is often seen in a negative light,
as something tawdry, deceitful, largely a battle between arrogant and
egotistical men and, worst of all, a waste of time. Nonetheless, a
'narrow' view of politics provides workable boundaries for political
enquiry.

Critics of this perspective see politics in broader terms, as far more
universal, capable of crossing cultural boundaries and existing within
and outside the institutional boundaries of the modern state (and
outside Europe!). Politics is not just confined to the actions of
government but is also found in the so-called private sector of the
business 'community' and in the more informal realms that often
operate outside the state. In fact, Leftwich argues that politics exists
'at every level and in every sphere' of human societies and that
political

activities are not isolated from other features of social life. They
everywhere influence and are influenced by, the distribution of power and
decision making, the systems of social organisation, culture and ideology

> in society, as well as its relations with the natural environment and other
> societies. Politics is therefore the defining characteristic of all human
> groups and always has been.
>
> (1983: 11)

Politics, in Leftwich's terms, occurs in homes, sporting clubs, work
places and in the street, as well as in parliaments. A strength of
Leftwich's inclusive depiction is its capacity to analyse non-institutional
politics. In this way, politics refers to our relationships to one another and
our interactions, in many different collective and sub-cultural forms: as
individuals; as members of families; as informal networks and groups; as
organisations; as governments; as corporations; and our activities in a
whole range of other institutionalised settings.

All definitions of politics are contested and value-laden. There is no one
right definition. It is important for students of politics to be explicit about
the reasons used to justify one definition over another. Either the narrow
or the broad definition of politics could be used to frame a consideration
of environmental politics, but quite different studies would be produced.
Use of the narrow definition would give an account based on the
constitutional design of the political process, its institutional
characteristics, and the role of political parties and pressure groups as
well as a consideration of the way in which environmental policy is made
and administered. And, indeed, these topics are covered in this book. A
broader definition would lead to the inclusion of non-institutionalised
forms of environmental concern and activism, which are never simply
captured by an organised mobilisation into normal politics. Attention is
also directed towards the political debates and conflicts that occur within
these informal movements and in the more non-institutionalised settings.
These, too, are included in this study. A narrow definition keeps more
things out of consideration, and hence makes the task of analysis more
manageable. A broader definition brings in more things that need to be
included but at the risk of a more diffuse and unmanageable study. There
are costs involved in the use of either definition: it is good to be aware of
what is gained or lost by choices made about preferred definitions.

What is the place of politics in environmental studies?

> Environmental studies' . . . essential interdisciplinarity is intellectually
> testing and touches not only on biology and ecology, but all the social
> sciences, most especially politics and economics.
>
> (Doyle and Walker 1996: 1)

It is tempting to say that politics is just one discipline that should be incorporated into environmental studies but far more is required than this: politics is central to environmental studies. The relationships between the two differ, depending on the definitions of each. Like politics, environmental studies is defined in numerous ways to include and exclude various modes of inquiry. Many of these differences emerge around conflicting views as to exactly what the environment entails. For example, if the environment is seen as a biophysical reality existing outside of humans, then environmental studies is the study of the relationships between human and non-human worlds. Often this 'external' view of the environment gives birth to instrumentalism; that is, the environment (or nature) is understood as a series of resources (natural products) or a series of processes, which need to be managed by humans in a renewable, sustainable, manner. Doyle and Walker argue as follows:

> . . . environmental studies is fundamental in the fullest sense. Human existence depends completely on the continuing availability of inputs such as air, water, foodstuffs and other resources from the natural environment. Without ecology there can be no economy and no society.
>
> (*ibid.*: 1)

This instrumental depiction of environmental studies fits snugly with many definitions of politics. For example, Leftwich's writings on politics portray the interface between humans and 'natural resources' as a centrally defining part of politics. He contends that

> the major organising activity at the heart of this history of cooperation, conflict, innovation, and adaptation in the use, production and distribution of resources has been and still is *politics*.
>
> (1983: 12)

As society is seen as outside the environment in this view, the study of politics complements the natural sciences (and the many other disciplines involved in environmental studies) in allowing us to understand and manage our biophysical resources; to manage our environment; to formulate and to implement environmental policies.

The role of politics in environmental studies increases in scope quite dramatically if we consider a different notion of the environment where humanity is part of nature and not distinct from it. Nature is no longer seen as a resource but as a more holistic construct that is far more open to interpretative possibilities. Being considered part of nature, there is a recognition that our relationships with the non-human world are socially, as well as (on occasions) biophysically constructed. So environmental studies now includes those relationships already mentioned between humans and non-humans, between differing non-human entities, and the relationships between humans themselves. Under this rubric, politics meets environmental studies in different ways. For example, issues of social democracy (participatory and representative), non-violence, social equity and justice as well as ecology now dominate the intellectual and green activist agenda. New types of relationship and entire human and non-human societies are analysed and imagined based on this more integrative, non-anthropocentric view of the environment.

Concepts for the analysis of environmental politics

Political regime and environmental politics

The particular way in which the political system is organised has a strong impact on the scope and effectiveness of environmental politics. There are a whole variety of different political regimes, and different ways in which rulers and ruled are connected and the institutions of politics are designed. For the purpose of this discussion, just three basic types will be described: (1) liberal-democratic representative; (2) military dictatorship; and (3) one-party authoritarian regimes. There are many variations within these three basic types but these will be sketched in only where necessary.

Examples of the liberal-democratic representative political regime are to be found in North America, Europe and Australasia, with variations in Japan, India and South Africa. Here the link between rulers and ruled is based on elections and the vote. Each citizen has a right to vote and a right to take part in politics. Periodic elections are held where political parties compete for the votes of citizens. On this basis, government is both formed and legitimated. Such a government is authorised to act on behalf of its citizens until such time as another election is required.

This can be characterised by the label 'liberal-democratic representative regime' because of the three main elements combined within the system of rule. First, there are liberal elements, characterised by a bundle of rights and freedoms that are necessary for participation in this type of politics. These rights typically include the freedom of speech and opinion, and freedom of assembly. Such freedoms can be embodied in a Bill of Rights (as in the USA), in a written constitution (as in France,

Germany and Italy), or in the conventions and practices known as the common law in states without written constitutions (as in the United Kingdom). Second, there is the democratic element symbolised by the right to vote in periodic elections; that is, the right to have a say or determine which political party or parties will form the government. It is in this limited sense that the people can be said to rule or govern themselves in this type of regime. Third, there is the representative element. These are not political regimes based on direct democracy where the people have a significant or effective say in all aspects of everyday governance. Instead, the people rule in an indirect manner, through elected representatives who take part in government, if their political party is in office and if they are selected to serve as ministers.

There are many different variations within political regimes of this type. Electoral systems can differ greatly from 'first-past-the-post' in single-member electorates, where the person who gets the most votes wins (as in the United Kingdom), to forms of proportional representation with multi-member electorates (as in Germany), and to preferential voting where the winner requires an absolute majority constructed by progressively eliminating the lowest-placed candidates and distributing preferential votes (as in Australia). Such states can take a unitary form (as in the United Kingdom) or some form of federation (as in Germany, the United States of America, India or Australia), where legislative powers and responsibilities are divided between a central or federal government and the states or provinces. Again, there are many different ways in which federal systems can be designed and operated. Party systems, the form of pressure group activity, and the role of trade unions, business groups and religious organisations can all vary widely. A comparative understanding of what happens in environmental politics needs to take into account and evaluate the significance of these variations.

The openness of liberal-democratic representative regimes contrasts sharply with the closed nature of politics in authoritarian states ruled by various forms of military dictatorships. Military governments come and go and at present can be found in countries as diverse as Nigeria, Burma and Indonesia. Characteristically, the military comes to power through a *coup d'état*, displacing a previous regime, which can have been either authoritarian or some form of democratic regime. Once in power these governments are likely to be repressive and to limit the role of opposition on environmental or any other grounds. Just because these are military regimes and rule in an authoritarian manner does not mean that politics comes to an end. The form of politics changes and the number of people who can be effectively involved is sharply reduced. The politics of opposition is also transformed so that much of its efforts to influence what is done is tied to arguments about the political form itself, often in attempts to introduce more open and more democratic political processes. For example, this can be seen in Burma with the struggle of Aung San Suu Kyi and her supporters against a harshly repressive military regime. Within the military there will be the standard jockeying for power and, at times, efforts at internal reform.

There is another sense in which politics continues inside these military regimes and it is surprising how often forms of representational politics are mimicked in these regimes. For example, almost as soon as the Suharto regime came to power in Indonesia, it set up a political 'party' and various trade unions, business organisations and civic bodies to try to integrate the population into the regime. Opposition political parties exist, even if their leadership can be determined by the government, and elections do take place. A similar process occurred in Chile when the Pinochet regime came to power after destroying the elected Allende government. These 'representative' organisations and processes provide some scope for the practice of oppositional politics, including a politics based on environmental concern.

Just as with liberal-democratic representational regimes, there is considerable variation between these military regimes. They differ in terms of the means by and the circumstances in which they come to power. They differ greatly in the amount of repression used to maintain themselves in power. There may be a great deal of difference between the development strategies produced by these governments, although frequently economic development will be linked to strategies for gaining wealth for particular individuals and families. They will also differ in the scope they give to the press and other opportunities for public debate. Finally, they will differ in the degree of stability and security of their domination.

There is another set of authoritarian regimes that can be characterised in terms of one-party domination. In the recent past, many of these were the classic one-party states like the USSR, the Peoples' Republic of China and the Democratic Republic of Vietnam. Here a party, claiming to be inspired by the works of Marx and Lenin, dominated the social, economic and political life of these countries, while vigorously pursuing economic development, often with a strong emphasis on rapid industrialisation. Once these regimes ran centralised, planned economies with rigid bureaucratic controls. Many of them now have 'liberalised' the economy but most still rigidly control the political process. Some have given up entirely and have been succeeded by forms of representative regimes of varying degrees of openness. Like authoritarian military regimes, these too mimic democratic and representational organisations. Elections are held but have more form than substance. Other 'civil' organisations lack independence and autonomy. Politics still happens, shaped here by the ideological claims of the regime and factional struggles. Such one-party regimes are not confined to the remnants of state socialism but can also be found in the Middle East. Examples include Qadhafi's Libya and Hussein's Iraq, as well as the contrasting regimes in Syria and Iran. One-party authoritarian regimes and military dictatorships often seem to show family resemblances as control and use of the military is an important part of the ways in which these regimes rule.

The character of the political regime has an impact on the scope and effectiveness of environmental politics. In liberal-democratic representative regimes there is significant scope for those concerned about environmental issues to have their say and to try to influence the political process. People are free to form groups, to join political parties, to invent new parties and to go into politics to achieve their goals. None of this says that environmental activists will be successful. Indeed, significant social and economic forces, equally able to organise, will oppose them and seek to keep environmental issues off the agenda. Nonetheless, the design of the regime provides scope for environmental politics. In authoritarian regimes, especially those committed to rapid economic development, there is very little scope for environmental politics, even by loyal supporters of the regime. Nonetheless, environmental politics and actions can still happen. The struggle of MOSOP (Movement for the Survival of Ogoni People) over the fate of the Ogoni lands in Nigeria included a strong environmental critique of the operation of the Shell Oil Company as part of its claim for autonomy. Environmental groups are active in Indonesia, despite frequent attempts at suppression. Even in Vietnam and China, some environmental issues are recognised in official circles.

Students of environmental studies need to be aware of these different kinds of political regime since they have an impact on the fate of environmental politics. They also need to be aware that there are other kinds of regime, other ways of linking rulers and ruled, and that these too have different consequences for the ways in which environmental politics can work.

Dimensions of power

Power is involved in all environmental conflicts and policy making. Both doing environmental harm and protecting, conserving or saving the environment require an effective deployment of power to prevail against opposition and power to make and enforce decisions. Unfortunately, the very terms needed to understand the character and consequences of power are contested. Both academics and political activists are divided in their views on what constitutes power, how power should be analysed or can be effectively deployed and how to assess the legitimacy of the ends of power in the different political systems around the globe. What is involved in this complex area of analysis can be illustrated by some relatively clear examples. Having done that it is then possible to consider the different ways in which power may be analysed and the ends of power evaluated.

Consider the problem of air pollution in the famous Italian city of Florence. From its ancient roots and its Renaissance pre-eminence through to its present, Florence has combined closely settled areas of housing and artisanal workshops with the grand monuments to its past. This has made Florence one of the great tourist centres of the world, but the number of tourists who flock to the city each year, coupled with a greater urban population, has meant an increased scope for environmental damage. Air pollution in the central area of the city is an intense and obvious problem as vast quantities of tourists, bus operators, normal cars and scooters rush their way through narrow streets, over the bridges in and through the city centre. Gradually and reluctantly, the city authorities have restricted traffic access to the centre. The problems of pollution have continued and a new campaign is under way to increase the zone subject to traffic restrictions. Some local residents and shopkeepers are openly hostile to these plans and campaign against them. Posters are stuck on many doors and in many shop windows claiming that the city (or their area) will die if the restrictions are imposed. Life will leave the city as shops close and its character will be irrevocably changed. On the other hand, the problem of air pollution is continually cited as a reason for action. Power operates on both sides of this debate. The local state, in trying to rectify a serious environmental problem, has to use power to have a regulation imposed and policed to get a change in a pattern of human and social behaviour with negative consequences for, in this case, an urban environment. The local residents are seeking forms of power that will stop this regulation being effective. Their means to power include argument, publicity, lobbying and demonstrations and may even include blockades to illustrate and press their claims. Here power to regulate and power to resist are counterposed.

Consider another kind of example. In the USA, major international mining companies have been involved in standard development versus environmental concern disputes. For example, in Wisconsin, Kennecott Copper, part of the global RTZ (Rio Tinto Zinc) enterprise, has sought to build an open-cut copper mine against the opposition of a coalition of Chippewa and white environmentalists (Gedicks 1993). The mining company has deployed considerable quantities of power, which are based on its wealth and its economic importance, to press its case and to marginalise its opponents. In pursuing its rights to business as usual in Wisconsin, as any company operating anywhere in the world would, Kennecott has used its power to mobilise investment (loans and shares), workers, technology and political access and the ability to influence popular and elite opinion to protect its corporate goals. To succeed, the Chippewa and their environmental supporters have to find sufficient power resources to deploy to deny Kennecott Copper its normal expectations to mine and to profit from its mining. The proposal to mine is as much based on power as are the efforts of those who oppose mining and seek to have a project stopped or regulated.

When knowledge about the causes and consequences of ozone depletion reached a certain stage, it was possible for the United Nations Environment Programme and environmental activists to place concern for the consequences of the increased uses of CFCs (chlorofluorocarbons) on the political agenda. Once again, a whole array of economic, social and political forces sought resources to be used as power as the conflict over regulating the relevant industries intensified. The chemical companies concerned, in alliance with governments, clashed with rival companies and countries over whether an international protocol was needed to regulate the production and use of ozone-depleting substances (Benedict 1991). After much struggle and the invention of a plausible, commercially profitable substitute, the USA took the lead in pressing for international regulation, which, after complex negotiations and a number of summits, came into effect. Here can be seen companies making strategic calculations about the benefits that could be derived from continuing or ceasing the production of CFCs and government officials finding strategies to make more or less international agreements. And the whole process turned around power and the effective deployment of arguments and resources that could be made to produce the consequences being sought.

Now, these three examples do not exhaust all the kinds of resources that are invoked and deployed to produce power-like effects and outcomes. These are only illustrations to show that situations of environmental conflict and policy making all contain elements of power. What happens in any given environmental conflict is the result of the creation and the successful deployment of forms of power. Environmental regulation, environmental neglect, business as usual, politics as usual; all these involve the deployment and playing out of power. What needs to be considered now is how best to understand the patterns of power involved in these routine, repeated forms of environmental conflict and policy making.

The analysis of power and models of the policy process

There are a large number of different kinds of account of the character, distribution and effects of power in politics. The points of cleavage in these debates are not agreed but there is a broad division between accounts that treat power as a quantity or resource to be deployed and those that do not. Most of the literature uses the quantity/resource assumption and is treated first, but it should be kept in mind that there is a broad alternative that deals with the same matters but in quite a different analytical framework.

An initial, most useful account of the different approaches in the dominant tradition has been given by Steven Lukes (1974) in a relatively old, very brief, accessible and influential volume *Power: A Radical View*. In this book, Lukes argues that there are three broad ways in which power is analysed and each of these is based on finding new 'faces' of power, successively adding these to form a complete account of the character and consequences of power.

The first face of power was identified and celebrated by the American pluralists. Their position is exemplified in the works of Robert Dahl (1961; 1970) and Nelson Polsby (1963). Dahl and Polsby were caught up in the behavioural methodological revolution for the social sciences and emphasise changes in observable behaviour as the key indicator of the presence and distribution of power. Following an earlier argument about the provenance of power (Weber 1978: 52), Dahl and Polsby believe they can detect its presence when the wishes of one person or group can prevail over the wishes of others. In their approach the best situations for the study of power are those involving observable conflict between groups seeking to influence government policy making. Group A wants one thing, Group B another and the government acts. The extent to which A and B get what they want indicates both the presence of power and the pattern of its distribution between them. In simple terms: 'A has power over B to the extent that A can get B to do something that B would not otherwise do' (Dahl 1957: 203). Of course, this formula needs to be methodologically extended to encompass even the normal political processes, which are the focus of Dahl's attention.

The application of this methodology was relatively simple and remains attractive. Find a situation of policy conflict, say a dispute between the proponents of sand mining on Fraser Island (off the Queensland coast, Australia) and those opposed on environmental grounds. Establish their contrasting demands and the initial position of the relevant government. Consider the power resources brought to the conflict and the skill with which they are used. Identify the outcome in terms of a specific government decision or action. Find which group got most of what it wanted and plot the distribution of power accordingly. Having done so it is possible to analyse the part that power played in the particular dispute and its resolution. In this case, since the sand mining of Fraser Island was stopped, environmentalists used their power resources more effectively than the mining companies and their supporters, and won.

The core of this position is based on a distinctive methodology and there is no particular reason why the methodology should have become associated with claims about the social distribution of power, but it did. The development of the pluralist approach came out of an attempt to refute the claims made about the unequal distribution of power in US society by elite theorists who were critical of the workings of US democracy. In contrast to the elite approach, which discovered concentrations of power, these academics found that there were no significant concentrations of power. Indeed power was diffused throughout society and the institutions of American democracy were such that diffused power resources would eventually get all voices listened to, and included in, the political process and in having their grievances addressed if not resolved. Much of the criticism directed at these 'pluralist' interpretations turns around not the question of methodology but the blatant unreality of this account of the distribution of power and its effects.[1]

There is an assumption in the pluralist/behavioural approach about the character of politics and the political process. It is assumed that people go about living their lives in whatever ways they choose or have chosen for them and that politics is a specialist sphere of society that is removed from the arena of their everyday interactions. People become political, choose to make issues political, enter the political process as a result of some disruption of, or disaffection with, the normal. A problem or grievance arises or there is a changed perception which renders something unacceptable that was previously acceptable. In these circumstances, individuals may seek political redress or, if the problem is shared, individuals might form groups (variously labelled interest or pressure groups) to put pressure on politicians to respond to their demands. Their ability to get what they want is a product of their power resources, the effectiveness of their organisation, the skill with which strategies and tactics are designed and power resources are deployed and, of course, the effectiveness of their opponents. Depending on their needs, people may join or form political parties and seek electoral success to address their problems. What is important here is the basic model of politics arising from grievance or interests with organisation and political action as a plausible response, seeking redress.

Lukes develops his account by tracing the way in which the methodological assumptions of pluralism are gradually undermined. For him, the next stage in the argument and the next face of power is found by adding conditions that undermine the utility of behaviourist assumptions. Here he focuses on the work of Bachrach and Baratz (1962; 1963; 1970) with their emphasis on agenda setting and a 'non-decision-making' power. For Bachrach and Baratz, power can be shown to be present when the actions taken by a group effectively prevent an issue getting onto the political agenda and marginalise the grievances of sections of the population in such a way that their needs never become political issues. In their own work, questions of race in American cities are used to illustrate this dimension of power. There is also a very effective account of contrasting city responses to air pollution that uses this approach in Matthew Crenson, *The Un-Politics of Air Pollution* (1971). Here the political conditions that allowed air pollution to be placed on the political agenda in one city were contrasted with those in another city where the issue did not emerge, even though the 'objective condition', air pollution from steel plants, was shared. In a sense, the second face of power revealed in these arguments probes further the assumptions about politics established by the pluralists. Politics is still about grievance, but here attention is paid to the role of power in preventing a grievance from being given effective political expression.

For Lukes (1974: ch. 3) this approach does not go far enough; too much of the behavioural methodology is still left in place and a further face of power needs to be added to give a complete picture. This face of power is not revealed by considering conditions of conflict and measuring who wins most from the resolutions of government, or by considering the deployment of power to shape a political agenda to accept some and marginalise other issues. Here the emphasis is on the way in which the interests[2] that people have may be denied even when those concerned remain unaware that that is what is involved. Power here is so effective that the very wants and desires of individuals are shaped by power to serve the ends of others. Into this realm the behavioural methodology cannot reach, and Lukes produces an argument by the invention of a 'counterfactual' to establish whose interests are served by an exercise of power even when there is no opposition and no observable behavioural change. It is not that the conception of politics has been broadened here. Rather the process begun by consideration of 'non-decision making' and agenda setting is taken a stage further, to consider a kind of power that allows people's interests to be harmed without them being either aware of or able to formulate the grievance upon which overt political action would be based.

In constructing an analysis of power involved in environmental conflicts or environmental policy making it is possible to use any or all of these dimensions of power and the methodological arguments associated with them. Elements of each of these faces of power may be present or relevant to a particular case study. If, for example, you wanted to study how the US government came, during the Reagan presidency, to reduce the scope for regulation by the Environmental Protection Agency (EPA), it would be possible to use a pluralist/behaviouralist methodology to study the competing efforts of environmental and business groups to get the government to change its mind and act differently. It would be possible to use arguments about agenda shaping to see why and how power was used to construct an effective political agenda for deregulation that was able to marginalise the previous institutionalised form of environmental concern. Further, it would also be possible to extend the argument, by reference to Lukes' third face of power, to the impact that this had on the real interests of those affected by the decision to lessen the level of environmental regulation and surveillance.

All the above approaches can be treated together as part of a broad tradition in the analysis of power, covering liberal, conservative and Marxist positions. They share similar conceptions of politics as grievance-/interest-based and power as a capacity and quantity that determines what happens, who wins, who gains and who loses by what is done. There is (and perhaps always has been) an alternative conception of power that is based on a different set of assumptions about the analysis of society and politics. The best contemporary expression of this can be found in Barry Hindess, *Discourses of Power* (1996), which, inspired and informed by a critique of the work of Michel Foucault, provides a sharp response to the quantity conception of power.

Back when behaviouralism and pluralism were conducting analytical battles with the elitists, an alternative view of power was proposed by the structural functionalist Talcott Parsons (1957). Parsons produces quite a different conception of power, which displaces much of the emphasis on conflict and the explanation of events in terms of different quantities of power in the hands of interested (that is possessing interests) actors. His much-cited definition is as follows:

> Power then is generalized capacity to secure the performance of binding obligations by units in a system of collective organization when the obligations are legitimized with reference to their bearing on collective goals and where in case of recalcitrance there is a presumption of enforcement by negative situational sanctions – whatever the actual agency of that enforcement.
>
> (cited in Lukes, 1974: 27–8; Hindess, 1996: 34)

As others have observed, this ties power to conceptions of legitimacy and the pursuit of 'collective goals' as opposed to all other forms of action and conflict. This is not all. For Parsons, power is not just negative or coercive but, as he expresses it, a medium like money, one that enhances or increases the capacity to get things done collectively that would be difficult or not possible to achieve individually. It was on this basis that he rejected C. Wright Mills' analysis and the more general emphasis on inter-group conflict because it treated power as if it were only a 'zero-sum' game.

Parsons' approach is strongly conditioned by his broader theoretical conception of society and, although it enjoyed considerable prestige in its day, and a degree of imitation, it is not that influential now. It is also less easy to apply his insights to the analysis of either environmental conflict or policy making. Nonetheless, there are issues that can be illuminated by applying his basic position. The rise of environmental regulation can be seen as a case of an enhanced collective capacity/ability to act based upon the increased willingness of voters to treat such action as legitimate. Voter authorisation legitimated an increased capacity of government to act to define and secure collective goals over environmental policy regardless of any specific conflicts over particular issues.

Parsons is not the only author with this concept of power as being positive, in the sense that it supports some collective purpose, as opposed to power as negative, to constrain, limit or prevent other things being done. Arendt's work on violence (1970) also includes an argument about power as the product of legitimacy and processes of popular or group authorisation.

This line of reasoning, or at least elements of it, reappears in some of the work of Michel Foucault and in the commentary on the concept of governmentality developed by those influenced by his work. Foucault produced a large number of studies combining arguments about power, knowledge and different forms of social and personal life, ranging from prisons to sexuality. He also wrote a number of books that deal more directly with questions of method and gave many interviews reflecting on his method, the political implications of his writing and on the interpretation of his work as a whole. Within these it is possible to see the outlines of an alternative conception of power.[3] Part of this alternative are versions of the argument about the performative/productive character of power, the coming together of people, their combination, producing the circumstances for more and different things to be achieved. Much of it is a reconfigured version of the rejection of power as a zero-sum game, expressed as rejection of the 'repression hypothesis' and an overemphasis on the negative side of the presence and operation of power. Also involved is Foucault's continual insistence on the connection between power and modes of resistance, which advertise themselves as a resistance to power but which are another face or part of power strategies. Foucault's work on power and the construction of subjectivity serves also to undermine the 'standard' of personal or inner autonomy universally used to establish repression and the negative impact of power. Further, the many metaphors Foucault produces to describe the operation of power focus his account on the conception of power as strategic and interactive rather than just as a clash of quantities of power with the distribution of power read back mechanically from an assessment of outcomes, the view that dominates in the power debate described above.

Foucault's lecture on governmentality (1991) has also been very influential for those trying to systematise this alternative 'Foucauldian' interpretation of power. In 1978, Foucault gave a series of lectures that included his discussion of 'governmentality'. This one lecture found its way into English via an Italian transcription of the original, given in French. It is important to note that Foucault never used the term 'governmentality' in his published works, although a few passages of the text find their way into the *History of Sexuality*, Volume I. As such it was not an important expression of Foucault's interpretation of social life. Since then the conception of 'governmentality' has been linked to Bruno Latour's notion of 'governing at a distance' and has been applied to a whole range of policy areas to produce challenging and contested interpretations (Dean 1991; Curtis 1995).

At its core, 'governmentality' emphasises the importance of knowledge and modes of calculation as well as internalising these in various subjectivities to explain what happens in particular institutions and social settings. For example, Miller and Rose (1993), in their article 'Governing Economic Life', use the concept of governmentality to show how a change in an accounting process was generalised and internalised in the modes of calculating investment as part of an effort to improve the efficiency of the British economy. Miller and Rose go to some lengths to show that the source of this change was not 'government' or the 'state' as such, but a whole range of diverse sources, including the autonomous actions of accountants acting as a profession. Policy and policy outcomes in this view are as shaped by the ways in which 'problems' are conceived and measured as by the efforts by those in conflict seeking to invent and use power resources. For the study of environmental politics, conflict and policy making, an easy illustration of the reordering of the interpretation of power comes from a consideration of the operation of environmental impact assessment. Frequently, environmental impact assessment is treated as a form of government regulation imposed on business from the outside, by the state, and resisted by business. Considering the same process through the interpretative lens of governmentality, attention would be paid to the way in which the institutionalisation of environmental impact assessment is linked to the development of new ways of knowing and calculating (and managing) the consequences of economic activity. Studies would reveal that new expertise is generated in the enterprise, that managing structures and processes are modified and that the way in which firms consider their products and production procedures could be changed. Such an interpretation could be easily grafted onto the interpretation of the EIA process given by Bartlett (1990) and Schrecker (1990), which focuses on the way in which 'ecological rationality' (Dryzek 1990; 1987) can invade and transform the 'administrative mind' (Paehlke and Torgerson 1990). More work needs to be done on the way in which this alternative interpretation of power can be applied to environmental conflict as well as environmental policy making.

Conclusion

Environmental politics is as varied as the issues, the activists and the political systems in which it is practised. All kinds of political variables combine to condition what happens and its significance for particular environments. In terms of political regime it does make a difference if environmental movements seek to achieve their goals in liberal-democratic representative regimes or under various forms of authoritarian, military or one-party rule. It also makes a difference how the institutions of politics are designed and linked. These factors together make up the institutional setting within which environmental conflict and policy making take place.

To assess both the character and fate of environmental politics requires a knowledge of these institutional factors, their implications for environmental activism and the issues that are the basis for environmental mobilisation. More than that it requires an understanding of the different arguments about how the role of power in politics can be assessed. This chapter has outlined the key concepts and alternative arguments needed to produce an effective interpretation of environmental politics either in a single country or in a number of countries. The comparative approach always strengthens our ability to be precise about the factors that condition the outcome of environmental conflict and policy.

This chapter has concentrated on the institutional, structural and 'permissive' factors that surround environmental activism. It is now necessary to turn to the active element, the way in which different individuals and social forces respond to the environment as an issue.

Further reading

Crick, B. (1964) 'The Nature of Political Rule', *In Defence of Politics*, Allen Lane, London.

Hindess, B. (1996) *Discourses of Power: From Hobbes to Foucault*, Blackwell, Oxford.

Leftwich, A. (1983) *Redefining Politics: People, Resources and Power*, Methuen, London.

Lukes, S. (1974) *Power: A Radical View*, Macmillan, London.

2 Political theories and environmental conflict

- Environmental resistance and reform
- Sustainable development
- Paradigmatic change
- Radical environmentalism

Introduction

Having analysed the character of the political process and the nature of prevailing political regimes, it is now necessary to consider some examples of the way in which politics has been applied to the question of the environment. Initially it is useful to examine the ways in which some very significant groups and governments have refused to accept that environmental damage ought to be treated seriously and have resisted efforts to put environmental concerns on to the political agenda. The contrast can then be established with those governments that have been willing to incorporate degrees of environmental concern as add-ons to 'normal' politics, or have been willing to reform administrative procedures and institutional structures to bring environmental considerations inside the policy process. It is then necessary to explore the whole plethora of ideas and claims that have motivated environmental activism, prompted the development of the complex of environmental movements and fuelled political and policy conflict over environmental questions. This chapter provides the foundation for a subsequent consideration of the internal character and dynamics of the various environmental movements that now operate in most countries around the globe.

Resisting the environment as a political issue

From the time that environmental concern started to be expressed in the United States of America and in Europe, those committed to growth were quick to mock its seriousness and relevance to politics. In the USA, figures like Herman Khan were conspicuous in their efforts to make a polemical assault on the findings and arguments of the Club of Rome (Meadows *et al.* 1972) and promulgate an alternative, optimistic futurology. Others were eager to join in and mock the seriousness of both environmental issues and those who sought to do something about them. Significant among these was John Maddox, editor of the influential British science journal *Nature*. His book, *The Doomsday Syndrome* (1972), was a very good example of the style, using his faith in science to debunk various claims about resource depletion, population growth, the negative consequences of an overemphasis on economic growth, and the increased scientific and technological redesign of agriculture.

Throughout the 1970s and 1980s, as the levels of environmental concern grew and activists struggled to get environmental issues on to the policy agenda, a continual stream of scepticism was poured on this endeavour by certain kinds of economists. All were concerned with portraying growth as the answer to economic, environmental and social problems. Some promoted free-market or market-like solutions to questions of environmental management (North 1995). Certain themes are stressed in these accounts. First, most claim to have some level of sympathy with environmental concerns but reject extremism and calls to urgent action. It has to be noted that evidence of their levels of environmental concern is hard to come by but some, like Wilfred Beckerman, did try to promote the use of market instruments to deal with pollution (1990). Even when doing so it is open to interpretation whether he was motivated more by a concern for the consequences of pollution or revulsion at the thought that government might use environmental concern to impose regulation on the economy. A similar ambiguity is to be found in most of the 'free'-market advocacy on environmental policy.[1] For example, in Bennett and Block (1991), a joint Canadian–Australian production, more time is spent trying to refute the case for environmental concern than in trying to define free-market solutions to these environmental problems. Again, the real enemy would seem to be the prospect of state intervention.

This genre of literature, focused on hostility to the consequences of environmental concern, enjoyed its prime in the early 1990s, by which time its themes had become quite predictable (Beckerman 1995; North

1995). There is no need for anxiety about population levels since human ingenuity and biotechnology will provide the food needed for whatever future levels of population there are; concerns about organochlorides are exaggerated, the product of chemiphobia, and the benefits outweigh the costs; the science of the enhanced greenhouse effect is complex and uncertain and there is no need for precipitate action (or any other kind, except research); the level of species extinction is exaggerated and the case for biodiversity is weak (besides, how many whales do we need anyway?); there is no problem with finite resources since human ingenuity, technology and price responses to increased scarcity will provide. Those who think otherwise are described as sincere but misguided extremists, zealots and Cassandras, undermining a proper and balanced concern with environmental problems and a recognition of the virtues of economic growth (North 1995).

The case for environmental scepticism wrapped up in free-market policy nostrums has not been without significant political consequences. In the United States of America, Ronald Reagan campaigned to emasculate environmental regulation in the name of free-market, pro-business reform. As a result, his head of the Environmental Protection Agency was cited for contempt of Congress and the head of Superfund (a scheme for cleaning up contaminated waste sites) served a term in jail for trying to enact his vision of a regulation-free environmental policy. This influence lingered on, symbolised by Bush's refusal to sign key protocols of the Rio Earth Summit and Clinton's willingness to compromise a whole range of environmental measures, all in the name of economic efficiency and the global competitiveness of the US economy.

Proposing reform

Politics with a green tinge

Given the success of environmental movements in having environmental issues put on to the agenda, it took only a little while for the political system to respond, even if that response was little more than just adding a tinge of green to the justification of politics as normal. In some cases, much more was done to institutionalise environmental concern than that.

The shallowest response can be seen in the statements of governments like that of Margaret Thatcher's in the late 1980s. After dramatic evidence of the consequences of pollution in the North Sea, Margaret Thatcher made a bold commitment to incorporate environmental concern into the policy making of her government. There was supposed to be a statement of national environmental goals with regular reports from government departments on how these goals were being met. A public relations flourish accompanied the announcement of these changes, but it was not long before the normal calculations of the Thatcher regime reasserted themselves. By 1996 it was recognised that the processes set up had been a waste of time as they were not being used to shape government policy at all. In this case, the only greening of politics came in the rhetoric used to describe what was already being done and to marginalise those who sought greater commitment to policy change.

Seeking sustainable development

Dressing up existing policy as if it contained green initiatives was not the only way of reaching an accommodation with rising levels of environmental concern. In this context and with increased tension between the developed capitalist countries and those poorer countries seeking to emulate them (symbolised by the debates over UNCTAD (United Nations Conference on Trade and Development) and the Brandt Report), a great deal of effort has gone into producing a concept of sustainable development that sought to combine the virtues of environmental concern with the pursuit of economic development and growth. These efforts have largely come from work sponsored by the United Nations Environment Programme and the Stockholm and Rio 'Earth' Summits. At Stockholm in 1972 the concept of sustainable development was only just being hinted at (Ward and Dubos 1972). It acquired some definition with the publication of the *World Conservation Strategy* (IUCN 1980).[2] This was followed up by some clarification and redrafting of various national conservation strategies (in Australia, the UK and the USA). All this was subsumed in slightly more detailed accounts in the Brundtland Report, *Our Common Future* (WCED 1987)[3] and the debates that flourished surrounding the Rio Earth Summit and the publication of *Agenda 21* (1992).

Although the arguments generated to link the concerns of economic development and environmental concern have varied over time (McEachern 1993), there is a core set of propositions associated with the concept of sustainable development. At its heart is the claim that the scope for sustainable economic growth is linked to the survival of sustainable environments. The flow of natural resources needed for economic production, the capacity of the soil to sustain food production, and the health of air, rivers and oceans, are all required to be able to imagine sustainable and repeatable economic growth and development. The slogan 'sustainable development is development that meets the needs of today's generation while not impairing the needs of future generations' captures much of what is at stake in the invention of the claim to shape policy by the requirements of sustainable development.

It should be noted that sustainable development is not a radical environmental or green concept, since it accepts the prime need for economic growth and the dominance of human welfare over the needs of the environment; and it conceives the relationship between humans and nature in terms of the use of the environment by and for humans. Nonetheless, adopting sustainable development does require a substantial rethinking of the terms of policy calculation and policy making and has at times been vigorously opposed by business and economists (Beckerman 1995: ch. 9), although many businesses have been willing to adopt the term as a best defence of their actions against environmental criticism. Further, the adoption of a sustainable development policy framework has sometimes been associated with political strategies to deal with the political consequences of rising (unpredictable) environmental concern.

The best example of this comes from the efforts of the Hawke Labor Government in Australia (1983–1991) to co-opt environmental concern to its electoral survival by seeking to codify the policy settings for ecologically sustainable development (ESD). Australia adopted sustainable development through a consensus-building process that concluded with the issuing of the National Conservation Strategy for Australia (1982). However, this document was not treated seriously and business adopted sustainable development to defend its existing practices from environmental criticism. Levels of politicised environmental concern intensified and the government became involved in a number of deft political manoeuvres, trading off environmental protection of specific sites for endorsement from environmental organisations in increasingly close elections. As a final twist in this saga, the government sought to incorporate environmentalists by involving them in the drawing up of new ESD policy frameworks. This time the focus was on broad policy parameters for the key sectors of the economy (such as manufacturing, mining and agriculture) and key intersectoral issues (such as urban systems and the response to the greenhouse effect). Drafting the ESD (1991) reports was the task of several working parties made up largely of bureaucrats with smaller numbers of business figures, environmentalists and trade unionists. No effort was made to find a consensus, although there seems to have been wide agreement over what should be in the recommendations. ESD turned out to be a more explicit sectoral version of sustainable development, characterised by a strong commitment to market environmental economics and certain minimal rules for assessing the environmental consequences of economic development. Rather like sustainable development, ESD was used as an indicator of the government's environmental concern but it did not interfere with its continuing promotion of economic development or the speeding up of project approvals (by short-circuiting the kinds of environmental baseline studies suggested in the ESD reports). Although ESD never became important for policy making, the reports do provide a good starting point for considering what sustainable development would look like if any government ever seriously wanted to implement it.

Radical environmental critique

There is little radicalism in the position of those who advocate either resistance or accommodation to environmental concern. Their actions combine to promote, at most, a slow pace of environmental reform with a great emphasis on broad policy statements, policy documents and the creation of 'new' institutions to look after environmental issues. Very few of these initiatives, with the possible exception of those sponsored by the United Nations Environment Programme, came from an environmental awareness that was not shaped by a response to a vigorous environmental politics generated by far more radical environmental movements. Environmentalists involved in these movements argue that green incremental change to 'business as usual' is not enough. They demand more deep-seated, widespread change in order to attain their environmental objectives. To understand the wide variety of these radical environmental positions, it is useful to consider the following five positions: deep ecology; social ecology; eco-socialism; ecological post-modernism; and eco-feminism. It is not possible, in the space available, to do much more than sketch the main features of these different arguments. There are very good surveys and arguments about green political theory that can be consulted for further details (Dobson 1995; Eckersley 1992; Goodwin 1992; Merchant, 1992).

Plate 2 *Buddhist temple, Inner Mongolia, China. Courtesy of M. A. J. Williams, private collection*

Most of these eco-radical philosophies propose paradigmatic change. Thomas Kuhn, in his immensely influential work *The Structure of Scientific Revolutions* (1969), described how bodies of scientific knowledge are created. Contrary to popular belief, Kuhn argued, those ideas that combined to make up 'best science' were not necessarily the best ideas available at any one time; rather they were the ideas that 'belonged' to the most powerful group of scientists. Kuhn referred to this constellation of ideas, beliefs and values as the dominant paradigm. This paradigm of knowledge could not accommodate new ideas sufficiently, as traditional knowledge was defended fiercely by its advocates. Consequently, revolutions in science necessarily occurred in cycles, replacing dominant paradigms with a completely new and radical constellation.

Radical environmental political theorists are involved in paradigm struggles, each seeking to create new sets of key values and principles that directly challenge existing, powerful paradigms. Each theory portrays the environment in a state of crisis due to the dominance of these powerful 'ways of seeing'. Some of these 'new values' are shared by a number of eco-radical theories, although each theory argues its position differently. For example, the deep ecologists level the bulk of their criticism at the anthropocentrism (human-centredness) of the dominant paradigm. Social ecologists argue that hierarchy is the problem. The eco-socialists attack capitalist principles as the main culprit. Ecological post-modernists identify modernity as the paradigm that must be challenged and overthrown. Finally, certain types of eco-feminists argue that the paradigm of patriarchy is largely responsible for widespread social and environmental destruction.[4]

Deep ecology

Deep ecology was a term first coined by Arne Naess in the early 1970s (Naess and Rothenburg 1989). Since then, it has also been referred to as 'ecocentrism' (Eckersley 1992). The most powerful assumptions of deep ecology are fourfold. First, ecocentrists argue that all beings, humans and non-humans, possess intrinsic value. The assumption that nature is a 'resource' is 'an essential and hitherto unquestioned axiom of western history and the economic and technological systems woven into that history' (Hay and Haward 1988: 437–8). Consequently this intrinsic value argument radically challenges this prevailing notion.

> The impulse to defend the existential rights of wilderness in precedence over human-use rights has led to a spirited challenge to the most fundamental tenet of western civilisation, the belief that rights are strictly human categories, and that no countervailing *principle* exists to bar humanity from behaving in any way it deems fit towards the non-human world.
>
> (*ibid.*)

The second major ecocentric assumption is that all beings are of *equal value*: 'that there are no "higher" and "lower" life forms in Nature' (Matthews 1988: 10). Third, there is the central principle of interconnectedness. Eckersley explains:

> According to this picture of reality, the world is an intrinsically dynamic, interconnected web of relations in which there are no absolute discrete entities and no absolute dividing lines between the living and the nonliving, the animate and the inanimate, or the human and the nonhuman.
>
> (1992: 49)

Finally, ecocentrists often argue that the Earth is finite in its carrying capacity, and that there are too many people on the planet: 'The flourishing of human life and cultures is compatible with a substantial decrease of the human population. The flourishing of non-human life requires such a decrease' (Sessions and Naess 1983). Deep ecologists are sometimes enthusiastic in their support of population control programmes. To deep ecologists then, human beings are not the unambiguous end or sole purpose of evolutionary progress but just another species existing on planet Earth (see Table 2.1).

These ecocentric arguments have been used in parts of North America, Scandinavia and Australia to campaign for, among other things, the preservation of wilderness. 'Wilderness' covers a vast range of associated subjects and approaches. It largely addresses the value of the non-human world or 'nature'. Wilderness can be valued instrumentally or intrinsically. Instrumental value is defined by humans for humans. Warrick Fox refers to one of these perspectives as 'resource preservation'. He explains it as follows:

> As the name implies, the resource *preservation* approach tends to stress the instrumental values that can be enjoyed by humans if they allow presently existing members or aspects of the non-human world to follow their own characteristic patterns of existence.
>
> (1990: 154, italics in original)

Table 2.1 *Key differences between dominant attitudes to the environment and those of deep ecology*

Dominant attitudes	Deep ecology
Domination over nature	Harmony with nature
Nature a resource, intrinsic value confined to humans	Natural environment valued for biocentric egalitarianism
Ample resources or substitutes	Earth supplies limited
Material economic growth a predominant goal	Non-material goals, especially self-realisation
Consumerism	Doing with enough/recycling
Competitive lifestyle	Co-operative lifeway
Centralised/urban-centred national focus	Decentralised/bioregional/ neighbourhood focus
Power structure hierarchical	Non-hierarchical/grassroots democracy
High technology	Appropriate technology

Source: Sylvan and Bennett 1986.

He lists nine separate types of argument that are used to justify the value of wilderness in human terms (*ibid*.: 154–61). These range from arguments about wilderness as a life support system for humans through to arguments about the aesthetic and religious experiences of wilderness. It is important to understand that all these justifications for preserving areas of wilderness and the wilderness experience are still anthropocentric. Deep ecology, on the other hand, is far more philosophically radical in its argument that non-human nature has an intrinsic value that is not dependent on acts of human valuation.

Again, there are many sub-categories of this position (*ibid*.: 162).[5] In this general case, humans are part of nature, as an equal component part with other species and natural entities. Here the theoretical separation between humans and the rest of nature is banished. Strategically, however, much of the emphases of ecocentrists have been less holistic: centred on those parts of nature that are less inclusive of people, the wilderness areas.

Social ecology

Social ecology is a position taken by a small but highly influential group within the broad spectrum of environmental movements. It has produced one of the fiercest critiques of the view of deep ecology and one of the most sustained points of opposition to the use of the state to provide solutions for environmental problems. Social ecology is a redesigned inheritor of the anarchist tradition with all its associated political values and hostility to the state, liberals and Marxists. Social ecologists stress the importance of human social organisation, advocate maximum individual autonomy and envisage society as a series of decentralised local communities, each strongly connected to a specific 'bioregion'.

Social ecologists argue that the marriage between ecology and anarchism is more than just one of convenience. They argue that there are fundamental similarities between the theories as they both identify some evolutionary telos that will deliver humanity and 'other nature' to some higher plain of existence. The fashionable catchwords of ecology normally reserved for describing 'healthy' non-human nature – interconnectedness, diversity, symbiosis, stability, flexibility and organicism – are seen as equally applicable to the anarchist vision of human societies (see Table 2.2).

Table 2.2 *Principles of social ecology/eco-anarchism*

1 Bypass/abolish modern nation-state. Confer maximum political/economic autonomy on decentralised local communities;

2 Centrality of political philosophy of anarchism. Close connections to ecology. Both draw inspiration from each other;

3 Opposed to all forms of human/non-human domination;

4 Strong defenders of grassroots/extra-parliamentary activities;

5 Consistency between ends and means.

Social ecologists are opposed to all forms of domination, both human and non-human. They advocate consistency between ends and means (i.e. political methods are just as important as their end result) and they are strong defenders of the non-institutional, grassroots politics of social movements (the details of which are outlined in Chapter 3). Just as Arne Naess is readily identified with deep ecology, so Murray Bookchin is the public face of social ecology or eco-anarchism. Like the work of many radical eco-political theorists, there is a notion of 'Ecotopia' in Bookchin's musings. He imagines a society where the principles of ecology and anarchism are one:

> If the foregoing attempts to mesh ecological and anarchist principles are ever achieved in practice, social life would yield a sensitive development of human and natural diversity, falling together into a well-balanced, harmonious unity. . . . Freed from an oppressive routine, from paralysing repressions and insecurities, from the burdens of toil and false needs, from the trammels of authority and irrational compulsion, the individual would finally be in a position, for the first time in history, to fully realize his potentialities as a member of the human community and the natural world.
>
> (1980: 187–94)

There are two key positions of Bookchin that need to be further considered: his theses on social hierarchy and evolutionary stewardship. Bookchin maintains that it is hierarchy in human societies that generates all forms of domination within human societies and between humans and non-human nature. Hierarchy, according to Bookchin, is a social construction, and has no 'natural' basis or essence.

> The domination of nature first arose within *society* as part of its institutionalization into gerontocracies that placed the young in varying

> degrees of servitude to the old and in patriarchies that placed women in
> varying degrees of servitude to men – not in any endeavour to 'control'
> nature or natural forces.
>
> (*ibid*.: 32, quoted in Eckersley 1992: 148)

Too often inequities within human societies are justified on the basis that
this unequal stratification exists 'in nature'. For example, conservative
and patriarchal aristocratic systems have been justified using the example
of the lion pride, with the most powerful male being classified as king.
But it is just as easy to project another model of politics on to the lion
pride. Feminist ecologists often view the dominant male in the lion
community as nothing more than a 'toy boy' who is used for seven or
eight months for breeding purposes, then traded in for a new one. The
point that Bookchin makes is that these hierarchies are human constructs
which are projected on to 'other' nature to justify differences in power as
somehow naturally essential within human societies. In this way,
Bookchin contends, human agency is denied in its bid to shape its polity,
and hierarchical divisions based on gender, class and race remain
ordained as 'naturally occurring'. He further suggests that once all forms
of hierarchy and domination are removed from human societies, then the
separation between humans and other parts of nature will also dissipate.

Bookchin's evolutionary stewardship thesis maintains that although
humans are part of nature (unlike the traditional resource
exploitation/conservation position), they occupy a rather more exalted
place in the evolutionary scheme (unlike the deep ecologists' 'equality of
all species' view). Indeed, he argues that humans are 'nature rendered
self-conscious' (Eckersley 1992: 155). Obviously there are some
similarities but many tensions between deep and social ecology, even
though both are radical green political theories (see Box 2.1).

Eco-socialism

> Eco-socialism is anthropocentric (though not in the capitalist-technocratic
> sense) and humanist. It rejects the bioethic and nature mystification, and
> any anti-humanism that these may spawn. . . . Thus alienation from nature
> is separation from part of *ourselves*. It can be overcome by
> reappropriating collective control over our relationship with nature, via
> common ownership of the means of production: for production is at the
> centre of our relationship with nature even if it is not the whole of that
> relationship. . . .

Box 2.1

Deep ecology versus social ecology

The 'battle' between social and deep ecologists came to a head in the late 1980s. In a series of publications, Murray Bookchin attacked deep ecology for its anti-humanism and lack of social perspective. Whilst deep ecologists sought to protect wilderness, Bookchin was demanding changes in the workplace, domestic and other human environments. He portrayed deep ecologists as 'deep Malthusians' due to their support for human population control (Bookchin 1987: 241). The following two extracts illustrate this tension. The first quotation is based on the words of deep ecologist and then Earth First! leader Dave Foreman; and the second is a response from Bookchin.

FOREMAN: When I tell people how the worst thing we could do in Ethiopia is to give aid – the best thing would be just to let nature seek its own balance, then let the people there just starve – they think this is monstrous. . . . Likewise, letting the USA be an overflow valve for problems in Latin America is not solving a thing. It's just putting more pressure on the resources we have in the USA.

(Foreman, n.d.: 43)

BOOKCHIN: It is easy to forget that it was out of this kind of crude eco-brutalism that a Hitler, in the name of 'population control', with its racial orientation, fashioned theories of blood and soil that led to the transport of millions of people to murder-camps like Auschwitz. The same eco-brutalism now reappears a half-century later among self-professed 'deep ecologists'.

'Deep Ecology', despite all its social rhetoric, has virtually no real sense that our ecological problems have their ultimate roots in society and in social problems. It preaches a gospel of a kind of 'original sin' that accurses a vague species called 'Humanity'.

This vague, undifferentiated 'Humanity' is essentially seen as an ugly 'anthropocentric' thing – presumably, a malignant product of natural evolution – that is 'overpopulating' the planet, 'devouring' its resources, destroying its wildlife and the biosphere – this, as though some vague domain called 'Nature' stands opposed to a constellation of non-natural things called 'Human Beings' with their 'Technology', 'Minds', 'Society' etc.

(Bookchin 1987: 221).

Many, more mild-mannered ecocentrist writers like Robyn Eckersley have attempted to play down this rift in radical eco-philosophy (Eckersley 1992: 146–60) and, indeed, there has been some reconciliation between the two antagonists quoted above. In 1991 Bookchin and Foreman published 'Looking for Common Ground', *Defending the Earth: A Dialogue between Murray Bookchin and Dave Foreman*, as part of this public reconciliation. In short, Bookchin now acknowledges the need to preserve wilderness, and Foreman has weakened his anti-humanist line arguing that there is a need to resolve social inequities affecting human population.

> Eco-socialism defines 'the environment' and environmental issues
> widely, to include the concerns of most people. They are urban based so
> their environmental problems include street violence, vehicle pollution
> and accidents, inner-city decay, lack of social services, loss of community
> and access to countryside, health and safety at work and, most important,
> unemployment and poverty.
>
> (Pepper 1993: 232–4, italics in original)

Like the anarchist-informed social ecologists, the eco-socialists (or
socialist ecologists/environmentalists) attempt to mesh ecological
principles with those of another, more traditional set of political theories
revolving around Marxism. In this instance, its ecotopian visions very
closely match that of certain types of socialism. Both social and socialist
ecologists are anthropocentric enough to believe that they should move
strategically from issues of social justice to ecology, not vice versa. In
this vein, both social and socialist ecologists fundamentally oppose deep
ecology. There are other similarities between the eco-anarchist and
Marxist environmentalists (sometimes referred to as the Green–Greens
and the Red–Greens) but as with all environmental political theories,
there are also many differences (see Table 2.3).

There are many forms of socialism and communism so, again, this
discussion gives only a broad overview. Socialism, as a political
tradition, pre-dates its reformulation by Marx and Engels in the
nineteenth century, but it was the Marxist version that was most
influential, having inspired the state-centred versions of both the USSR
and China, as well as the radical left critics of social democracy in the
liberal democracies. Socialism is focused on both social inequality and
the consequences of a class-divided capitalist society. In their famous
pamphlet *The Communist Manifesto*, Marx and Engels claimed:

> The history of all hitherto existing society is the history of class struggle.
>
> (1848)

Marx's writings were an elaborate and, at times, savage critique of
capitalism. He saw capitalism as creating two great, opposing classes in
society: the bourgeoisie and the proletariat. The bourgeoisie owned the
basic 'means of production' such as factories, machinery and the income-
producing property. The proletariat did not, and was totally dependent
on the only commodity it possessed, its ability to work, or in Marx's
analytical vocabulary, its 'labour power'. As a consequence of their
different relationship to the means of production, the class of labour had
to sell their labour power to capital. This allowed production and

Table 2.3   *Eco-socialism: some socialist–anarchist differences*

Socialism	Anarchism
Social injustice, environmental degradation caused by class exploitation	Social injustice, environmental degradation caused by hierarchical power relations
Class is defined by economic criteria	Class is also defined by non-economic criteria (race, sex)
Explanations and analyses are historical	Explanations and analyses tend to be ahistorical
Ambiguity about the state – favours at least localised forms of it	Total opposition to the state
Abolish capitalism first and the centralised state will wither away, because capitalism creates the state	Abolish the state first, as an independent act from abolishing capitalism, because the state creates capitalism
The state is the representative and defender of the bourgeoisie	The state represents its own interests, independently of other economic classes
Participation in conventional politics is permissible in the path to revolution	No participation in conventional politics is permissible
Revolution by subverting and confronting capitalism – experimental communities etc. are naive and Utopian	Revolution through by-passing capitalism and creating 'prefigurations' of the desired society, such as alternative communities and economies
Emphasise strength of collective political action	Tend to emphasise personal-is-political maxim and individual lifestyle reform
Revolution particularly via our collective power as producers, i.e. unions withdrawing labour, especially in a general strike	Syndicalists advocate union organisation and action – other anarchists stress civil disobedience by community and other non-economically defined groups
Working class will be major actors in social change	New social movements and community groups will be major actors in social change
Tendency to vanguardism (Marxism–Leninism)	No revolutionary vanguards
Tendency to dictatorship of the proletariat (transitional stage)	Any 'dictatorship' or government is anathema

continued . . .

Table 2.3 continued

Socialism	Anarchism
Materialist philosophy and approach to social analysis	Tendency to idealism
Modernist politics	Tendency to 'post-modernist' politics
Need for a planned economy	Communes should self-organise, within limits, because spontaneity is important
Limited support for decentralisation	Decentralisation is vital
Individual freedom may be circumscribed by the collective	Individual autonomy is vital
Necessary international exchange, based on reciprocity, is an important aspect of international socialism	Opposed to most international trade. Applauds local self-sufficiency
Ambiguity about a money economy. Only a few oppose	Most oppose a money economy
Urban-centred	Has a prominent anti-urban element, as well as urban anarchism
Conceives of nature as socially constructed	Tends to see nature as external to society but the latter should conform to nature's laws and regard nature as a template
Anthropocentric (but not in the same way as capitalism–technocentrism)	Advocates (in social ecology) neither anthropocentrism nor biocentrism
Advocates socialist development	Advocates various development models, inc. socialism, environmental determinism and independent development (bioregionalism)
Deep structures (especially economic) condition surface structures, such as spatial organisation	Spatial organisation is a determinant of economics, society, politics

Source: Pepper 1993.

consumption to occur but defined the relations between the classes in terms of a fundamental antagonism. On the basis of this structured opposition and the experience of exploitation, class conflict would be continuously generated. Marx saw the conflict between classes as the 'motor of history'. Inevitably, capitalist development would produce 'wealth at one pole, misery at the other', and an increasingly polarised society. It was Marx's expectation that sharp class antagonism would provide the basis for radical political change when the proletariat united to overthrow the social, economic and political power of the bourgeoisie and usher in a new stage in world history, socialism followed by communism.

For Marx, poverty was a result of capitalism, not the fault of poor people producing too many children. On this basis, Marx developed his own distinctive theory of population. Marx rejected Malthus' simple claim that the rate of population growth would inevitably outstrip society's capacity to provide all but misery and famine. For Marx, the population question was a sub-set of the relationship between expanding production capacities, capital accumulation and the ability of labour to defend its position. In some circumstances, the increase in population was a strategic response to the conditions of capital accumulation, where the expansion of production was dependent on a greater use of labour power. In other circumstances, the 'increase of labour power itself' could mean something other than just an increase in the size of the working population. Marx rejected the reactionary, and pessimistic, views of Malthus and criticised their crude restatement of certain class positions. He was more optimistic about the development of productive forces being able to provide and alter the dynamic of population growth. It is for this reason that Marx is seen as being too sanguine about the prospects of unlimited growth without population or resource constraints; hence, resource security was treated as a consequence of bourgeois exploitation and capitalist overconsumption rather than of natural limits coupled with overpopulation.

Obviously, Marx's anti-Malthusian sentiments are similar to Bookchin's critique of deep ecology, but in this case, hierarchies of class, gender and race are replaced by classes defined purely by their position in the processes of commodity production.

Environmentalist critics of this socialist position argue that this inability to accept 'natural limits' must mean that Marx's writings cannot be reconciled with an ecocentric perspective. Bell comments:

> Exhortation that socialism must ultimately come to grips with the
> question of ecological constraints on growth is one thing, but in general
> socialists have certainly not seized on this issue. The dominant socialist
> position would still be that a socialist transition, by overturning
> capitalism's inefficiencies and waste, could spur the growth process:
> outdoing capitalism at its own game.
>
> (1986: 11)

Some scholars argue that too much emphasis has been placed on this
point since Marx did not overtly object to the notion that human
population numbers and natural resource limits are linked. He did object,
however, to the conservative values smuggled into the Malthusian
'principle of population', which 'dictated that the axe of subsistence
should fall on the necks of the poor rather than the rich' (Doyle and
Kellow 1995: 45).

Whatever the nuances, perhaps this is one socialist principle that has
been modified to make for the development of an eco-socialist position.
In his defining work on eco-socialism, David Pepper writes:

> The eco-socialist response to resource questions is not merely to fix on
> distribution, as commentators like Eckersley (an ecocentrist) suggest. It
> says that there are no ahistorical limits of immediate significance to
> human growth as *socialist* development. But there are ultimate natural
> constraints which form the boundaries of human transformational
> power.
>
> (1993: 233, italics in original)

Ecological post-modernism

The environmental movement has always included those who have
rejected the whole social project of the Enlightenment. These people see
environmental damage as a product of 'Enlightenment thinking', which
extolled the virtues of rationality and the ability of humans to use
'nature' for their own ends in the name of progress and an ever-
increasing standard of living. Technology and the whole project of
modernity is seen to have produced as many problems as it has solved
and the claims of progress are fairly hollow. For these critics, the final
achievement of progress and reason will be ecological destruction.

For much of the time, these views could be called anti-modernist, since
they define their position in terms of a rejection of modernity. These anti-
modernist views have produced some of the most enduring images of

contemporary environmental protests, since it was the anti-modernists who rejected the codes and practices of affluent lifestyles and sought simpler ways of living, in tune with nature. In many places, these people left cities, sought out the countryside and the forest, and mounted their critiques of the overdevelopment of the cities.

In recent years, there has been an accommodation between these anti-modernist views and the post-modernism and post-structuralism of the academy (Cheney 1989). Post-modernism, in simple terms, sees Western societies as transformed by modernism to such an extent that modernist principles and beliefs no longer hold and, through excess, turn into their opposite. Here the link can be made: post-modernists too see the products of rearranged technological processes delivering increasingly negative consequences for the Earth and its populations. Some of these post-modernists embrace a dethroning of reason and a more fragmented, decentralised social world.

Post-modernists tend to emphasise the importance of locality and difference as against centralisation and the notion of an homogenised sameness. Post-modernists and post-structuralists would, on the whole, be sceptical of claims about an 'essential nature'. Nature itself would be seen as a social construct and the relations between humans and nature would be capable of almost infinite variety.

Many are extremely hostile to this environmentalist critique of modernity and Western science. Martin Lewis writes:

> In one sense, the romance between ecoradical philosophy and post-modernism is entirely natural. Postmodernism began its career with an attack on the sterile architecture of high modernism, and what, after all, could be more antithetical to the organic ideal of the greens than Bauhaus 'machines for living'? Similarly, the poststructuralist offensive against the supposedly disembodied, logocentric, objectivist, totalizing, imperialistic, Eurocentric project of Western science and rationality conforms impeccably with the ecoradical critique of Western philosophy. Concordance can also be found in the emphases placed on diversity in cultural expression and in nature itself. Even the postmodernists' accent on playfulness finds its echoes in the Greens' favoured account of natural processes.
>
> (1996: 219).

Much of Lewis' depiction of eco-radicalism is based on an attack on deep ecology, since deep ecologists often share this disdain for Western science with their ecological post-modernist counterparts. There are, however, major differences between them. Whereas deep ecologists see 'nature' as some transcendental force with intrinsic value, ecological post-modernists would deconstruct this depiction of nature and treat it as a social construction with inventing 'value', intrinsic or otherwise, a fairly standard human pretension.

Eco-feminism

Eco-feminism emerged in the 1970s as part of the women's liberation movement. Carlassare (1994) argues that the term 'eco-feminism' is utilised 'by some activists and academics to refer to a feminism that connects ecological degradation and the oppression of women'.

The women's movement remains one of the most vibrant movements advocating social change at the turn of the twentieth century. Like the environmental movement, feminism is full of important and different perspectives, far too many to do justice to here. Caroline Merchant identifies four positions that can usefully be discussed: liberal feminism, Marxist feminism, cultural feminism and social feminism (see Table 2.4; Merchant 1992: 186–7). She writes:

> Liberal, cultural, social and socialist feminism have all been concerned with improving the human/nature relationship and each has contributed to an ecofeminist perspective in different ways. Liberal feminism is consistent with the objectives of reform environmentalism to alter human relations with nature from within existing structures of governance through the passage of new laws and regulation. Cultural ecofeminism analyzes environmental problems from within its critique of patriarchy and offers alternatives that could liberate both women and nature.
>
> (1992: 184)

Social eco-feminism is more closely attuned to the eco-anarchist writings of Bookchin. The similarities between the two are many, with emphases on decentralisation and attacks on all forms of hierarchy, particularly patriarchy.

> Social ecofeminism advocates the liberation of women through overturning economic and social hierarchies that turn all aspects of life into a market society that today even invades the womb.
>
> (*ibid.*: 194)

Finally, socialist eco-feminism is closely intertwined with socialist ecology (eco-socialism); but in this case, it is reproduction not production, 'which is central to the concept of a just, sustainable world' (*ibid*.: 195).

As there are many divisions between different radical environmental camps, so too these divisions manifest themselves within eco-feminism. One of the most interesting discussions revolves around the 'essentialist/ constructionist' tension (Carlassare 1994: 221). The essentialists, and some cultural feminists, argue that women are closer to nature than men, and this innate quality should be recognised, further developed and celebrated. Some liberal, social and socialist eco-feminists, of course, abhor this suggestion, arguing that social factors, not innate qualities, are seen as the most important shapers of gender-based inequalities (see Salleh 1992 and Cuomo 1992 for these more constructionist arguments).

Using the essentialist/constructionist axis, Lewis refers to the essentialist eco-feminists loosely as 'radical eco-feminists', while the constructionists are more 'mainstream'. He contends:

> To many feminist thinkers, the notion that female thought is fundamentally different from male thought in any way – especially by being more in tune with nature – is both offensive and dangerous. This line of reasoning . . . has long been used by men to keep women in subjugation. When women are identified with nature, men are able to appropriate the realms of science and rationality for themselves.
>
> (1992: 35; see also Plumwood 1988)

Vandana Shiva is another eco-feminist who attacks the essentialists, this time from a self-conscious third world perspective. She sees colonialism and patriarchy as explaining the success of the globalisation of 'Western development' and advanced capitalism. She explains:

> Development was thus reduced to a continuation of the process of colonization; it became an extension of the project of wealth creation in modern Western patriarchy's economic vision, which was based on the exploitation or exclusion of women (of the West and the non-West), on the exploitation and degradation of nature, and on the exploitation and erosion of other cultures.
>
> (1994: 273)

Shiva goes on to argue that this particular kind of oppression explains why women and other subjugated cultures have been direct activists in opposing modernisation and development in parts of the third world, 'struggling for liberation from development just as they earlier

Table 2.4   Feminism and the environment

	Nature	Human Nature	Feminist Critique of Environmentalism	Image of a Feminist Environmentalism
Liberal Feminism	Atoms Mind/body dualism Domination of nature	Rational agents Individualism Maximisation of self-interest	'Man and his environment' leaves out women	Women in natural resources and environmental sciences
Marxist Feminism	Transformation of nature by science and technology for human use. Domination of nature as a means to human freedom Nature is material basis of life: food, clothing, shelter, energy	Creation of human nature through mode of production, praxis Historically specific – not fixed Species nature of humans	Critique of capitalist control of resources and accumulation of goods and profits	Socialist society will use resources for good of all men and women Resources will be controlled by workers Environmental pollution could be minimal since no surpluses will be produced Environmental research by men and women

Table 2.4 continued

Cultural Feminism	Nature is spiritual and personal Conventional science and technology problematic because of their emphasis on domination	Biology is basic Humans are sexually reproducing bodies Sexed by biology/gendered by society	Unaware of interconnectedness of male domination of nature and women Male environmentalism retains hierarchy Insufficient attention to environmental threats to women's reproduction (chemicals, nuclear war)	Woman/nature both valorised and celebrated Reproductive freedom Against pornographic depictions of both women and nature Cultural eco-feminism
Socialist Feminism	Nature is material basis of life: food, clothing, shelter, energy Nature is socially and historically constructed Transformation of nature by production and reproduction	Human nature created through biology and praxis (sex, race, class, age) Historically specific and socially constructed	Leaves out nature as active and responsive Leaves out women's role in reproduction and reproduction as a category Systems approach is mechanistic not dialectical	Both nature and human production are active Centrality of biological and social reproduction Dialectic between production and reproduction Multi-level structural analysis Dialectical (not mechanical) systems Socialist eco-feminism

Source: Merchant 1992.

struggled for liberation from colonialism' (*ibid.*: 273; Mies and Shiva 1993).

Conclusion

So far no government has tried seriously to take environmental concern into the process of policy making. Specific environmental disputes and the occasional environmental issue may have prompted government action (such as the banning of CFCs, the saving of particular areas of forest, the creation of game reserves or the regulation of pollution) but mostly whatever has been done has been done to accommodate rising levels of environmental concern and to protect the demands of economic development and economic growth. The way in which the accommodation between growth and concern has been constructed has varied from the cosmetic to the opportunist, with rare attempts to develop and embed rules that cover the best way to manage the impact of economic development on environments. With strategies of incorporation and accommodation to the fore, there has been plenty of scope for more comprehensive environmental critiques to develop and more radical forms of environmental politics to arise as environmentalists see their concerns marginalised or ignored.

Further reading

Kuhn, T.S. (1969) *The Structure of Scientific Revolutions*, University of Chicago Press, Chicago.

McEachern, D. (1993) 'Environmental Policy in Australia 1981–1991: A Form of Corporatism?', *Australian Journal of Public Administration*, Vol. 52, No. 2: 173–86.

Pearce, D. (ed.) (1991) *Blueprint 2: Greening the World Economy*, Earthscan Publications, London.

3 Environmental politics in social movements

- Social movement theories
- Different environmental movements
- 'North' and 'South'
- Global ecology

Introduction

Environmentalism, in all its forms, was born in environmental movements. There are many theories about what makes up a social movement, and some of these are outlined in this chapter. At the outset, what needs to be understood is that social movements are largely non-institutional. They occupy a political terrain that is often quite separate from more established institutionalised political forms such as pressure groups, parties, and the administrative and parliamentary systems of the state. It was within these non-institutional, more informal realms of society and its politics that environmental movements emerged. It is safe to say that without the environmental movements there would be little or no 'greening' of government and corporations.

Comparatively little has been written about this more informal realm of politics. Traditional political science has largely ignored the politics of everyday life, doubting that it is important enough to warrant analysis. Social movement theorists, however, believe that most new and transforming ideas begin life in non-institutional politics. This creative 'politics of the people' is evident in dynamic, amorphous networks, associations, grass-roots groups and alliances (Doyle 1998). Rarely is this dimension governed by formal laws and statutes of association, such as constitutions.

The next chapter also includes an investigation of non-governmental organisations (NGOs) in environmental politics. While still treated as non-institutional politics in this work, some NGOs cross the non-institutional/institutional divide. These NGOs have legitimised themselves through the adoption of constitutions, setting rules of conduct and defining organisational goals. They are, in this sense, formal political institutions. To a degree, the adoption of a constitution symbolises that they are willing to work within established rules and social norms. Such NGOs are as formal as non-institutional politics gets. Considerable tensions surround the relationships of NGOs to the non-institutionalised grass-roots groups and networks of social movements on the one hand, and the institutions of the state on the other. For this work, NGOs are treated as being a constituent part of social movements, along with other sub-groupings such as networks and informal groups, rather than existing as an entirely separate phenomenon.

What are social movements?

Before it is possible to analyse the dynamic and diverse character of environmental movements, it is necessary to establish some of the general characteristics of social movements. 'Social movements' is a term used to refer to the form in which new combinations of people inject themselves into politics and challenge dominant ideas and a given constellation of power. The nineteenth-century labour movement is a good example. Here people who found themselves confronting harsh industrial working conditions joined together in a myriad of small organisations or combinations to press for changes, from their bosses and from the state. In Britain, for example, such 'combination' and 'oath taking' was illegal; workers did not have the right to organise or the right to vote, and defence of the rights of private property was of great importance to the state. Nonetheless, workers struggled for a vision of a better future, achieved their basic goals, transformed the rules upon which the system was run and ended up incorporated into the changed social order. The labour movement did not achieve its radical goals, or the overthrow of capitalism, but it did change the system to the extent that it and its concerns went from being excluded to being included.

In contemporary parlance, this labour movement would be seen as an example of an old social movement in contrast to the 'new' social movements, which include both the women's movement and the environmental movement. These are new in the sense that they challenge a new set of dominant ideas and another constellation of power. There are new issues and concerns to be injected into the political process. Like the preceding social movements they have a radical edge and visions of a world transformed by their demands. Their radicalism is heightened by their awareness of what happened before: that radical movements ended up being incorporated and their issues and passions tamed. New social movements are characterised by their informal modes of organisation; their attachment to changing values as a central part of their political challenge; their commitment to open and ultra-democratic, participating modes of organisation (at least in the initial stages); and their willingness to engage in direct action to stop outcomes that they see as harmful. It has certainly been the intention of these movements to disrupt the taken-for-granted routines of normal politics and to push other considerations to the fore.

The new social movements often take up themes left over from models constructed in earlier eras and give them new emphasis and meaning. Such was the case with the re-invigoration of feminist politics, the peace movement and anti-nuclear campaigns. In some ways a revised environmental politics was the same. Environmental campaigns had existed before, even in the early stages of industrialisation, and regulations had been imposed to limit pollution, largely on health grounds. From the 1970s onwards a revitalised environmental movement began working through a whole array of local and national networks and organisations to press its claims. Often there were direct actions, such as blockades, marches and rallies, as well as quieter attempts to lobby for policy changes and new initiatives. It is the whole array of these activities that is the subject of this and the next two chapters.

Why do social movements happen?

There are many theories about the origins and character of social movements. The most significant of these have their roots in the sociological theories of the Chicago School (Princen and Finger 1994: 48). Princen and Finger write:

> social movement theory goes back to psychosociology and the study of individual behaviour within groups. Collective action, according to this theory, can be triggered in various ways, depending essentially upon the theoretical framework to which one refers. One can distinguish three main schools. All of them are fundamentally ahistorical. Indeed, collective action can occur either as a result of relative deprivation, as a

> strategy to articulate common interests, or as a response to economic or political conflicts. In a political context the purpose of collective action is social change.
>
> (*ibid.*: 48–9)

In collective action theories, individuals are treated as rationally responding to forms of deprivation or to some newly presented opportunity for political success. Such situations usually occur when there is 'rapid social change' (Oberschall 1993: 18). In these circumstances traditional relationships and ideas are challenged, sometimes giving rise to social movements. Oberschall writes:

> In this view, a period of rapid social change – due to industrial growth and economic transformation, urban growth and rural decline, an economic depression, the aftermath of a lost war, rapid population growth, and the like – will weaken and undermine stable groups and communities. . . . As social bonds weaken and traditional answers and remedies no longer work, the population will manifest signs of increasing disorganization . . . they participate in major social, political, and religious movements that seek to reform and restructure institutions.
>
> (*ibid.*: 18)

Basically, these collective action theorists believe that something must go substantially 'wrong' for people to coalesce into new social movements (Box 3.1).

Box 3.1

Four factors prompting collective action

1   Changes in the basic conditions of life to produce discontent.

2   Changes in beliefs and values.

3   Changes in the capacity to act collectively.

4   Changes in opportunity for successful action.

(Source: adapted from Oberschall 1993: 17)

Collective action theories tend to be very general and ahistorical in their accounts of the development of either old or new social movements. In considering new social movements it is important to pay attention to the circumstances in which they arise and to the specific characteristics of both the participants and their goals. There are two broad explanations of this type, one that focuses on post-materialist values and one that focuses on the experiences of a post-industrial world.

One of the most significant and pervasive accounts of the origins of the new social movements has emphasised a 'value shift' in society explained in terms of a post-materialism thesis.

Environmental movements, for example, are seen as possessing post-materialist values that directly contest, in a paradigmatic battle, the dominant materialist values of modern society. This argument is commonly identified with the writings of Inglehart (1977, 1990; see also Papadakis 1993). Strongly premised on Maslow's 'hierarchy of needs' (1954), the post-materialist argument is that having largely fulfilled the more basic needs of safety and security, parts of advanced industrial society are able to pursue the 'higher', more luxuriant causes, such as love and a sense of belonging, beyond the old politics of material existence. Inglehart states:

> A process of intergenerational value change is gradually transforming the politics and cultural norms of advanced industrial societies. A shift from Materialist to Postmaterialist value priorities has brought new political issues to the center of the stage and provided much of the impetus for new political movements . . . from giving top priority to physical sustenance and safety toward heavier emphasis on belonging, self-expression, and the quality of life.
>
> (1990: 66)

It is accepted that some environmental movements do seek post-materialist values and express their politics in these terms. In parts of the more affluent world, arguments relating to the aesthetic values of nature, non-human rights, the spirituality of place, and an emphasis on holism and ecology would seem to fit the post-materialist hypothesis. It should be noted, however, that not all first world environmental movements are either predicated on or seeking post-materialist values. In addition, in poorer parts of the world environmental movements can be effectively based on those old survival/security needs in situations made worse by extreme environmental degradation. The struggles of the Ogoni people in Nigeria against pollution caused by Shell are a case in point. So, whatever the strengths and weaknesses of the post-materialist thesis, it can only explain a little about the origins and character of the environmental movement.

An alternative account of the origins of environmental movements is based on the thesis of post-industrialism.[1] This position argues that advanced industrialism, championed by both the market systems of latter-day capitalism and the state-centred models of Soviet-style socialism, has pushed the Earth, its habitats and its species (including humans) to the brink of extinction. This industrial/development paradigm has promoted economic growth at all costs. Initially this pursuit of growth was rooted deeply in the Enlightenment project of the Scientific and Industrial Revolutions, the pursuit of progress, and improved living standards for all. The environment, and nature, was presented as a cornucopia of unlimited resources and abundance. Initially, the environmental costs of growth were either not recognised or treated as incidentals in the gaining of a greater good. In more recent times, industrialism has become global and there is widespread (if partial) acceptance of natural constraints to growth, but the Enlightenment project continues. It still advocates increased growth but now this should be bolstered by improvements in environmental efficiency and management, the promotion of the global 'free market', and the advocacy of homogenous 'democratic', pluralist political systems. Carl Boggs writes from the post-industrialist perspective:

> to the degree that the radicalism of new social movements tends to flow from the deep crisis of industrial society, its roots are generally indigenous and organic, making it naturally resistant to totalistic ideologies that galvanized the Second and Third Internationals . . . the eclipse of the industrial growth model, the threat of nuclear catastrophe, bureaucratization, destruction of natural habitat, social anomie – cannot be expected to disappear simply through the good intentions of political leaders.
>
> (1986: 23)

For writers like Boggs, the post-industrial setting generates a unique social and political climate that promotes the formation of new social movements (NSMs). As a result, the defining characteristics of these movements differ from those that went before in terms of class, and ideological and organisational characteristics. (These differences have been summarised in Box 3.2.)

Boggs argues that NSMs are less likely to be co-opted than movements existing prior to the 1970s, although it is difficult to see why this is more than a hope on his part. Post-industrialist theorists, however, argue that current problems are so profound that they cannot be routinised into normal politics. It is true that some parts of green movements retain their opposition to dominant institutions. But it is also true that other parts have been readily co-opted and many co-operate with government as a

---

Box 3.2

Eight characteristics of new social movements

1 NSMs do not bear a clear relation to the structural roles of the participants. There is a tendency in NSMs to transcend class structure. More important are the different social strata provided by youth, gender, sexual orientation and professions.

2 Ideologically, NSMs are profoundly different from the Marxist perception of ideology as a unifying and totalising element for collective action. NSMs exhibit a multitude of ideas and values.

3 NSMs often involve the emergence of new dimensions of identity. The grievances are based on a set of beliefs, symbols, values and meanings, rather than on the economic grievances that characterised the working-class movements.

4 The relationship between the individual and the collective is blurred. Many contemporary movements are 'acted out' in individual actions rather than through or among mobilised groups. The movement becomes the focus for the individual's definition of himself or herself, and action within the movement is a complex mix of the collective and individual confirmations of identity.

5 NSMs often involve personal and intimate aspects of human life, e.g. what we eat, wear and enjoy.

6 NSMs use the radical mobilisation tactics of resistance, which differ from those practised in working-class movements, characterised by civil disobedience and non-violence.

7 There is disdain on behalf of NSMs for conventional politics. Consequently NSMs maintain elements of autonomy from traditional mass parties.

8 NSMs seem to be segmented, diffuse and decentralised. There is a tendency toward considerable local autonomy of local sections. (This point is developed at some length in the next section devoted to the 'structure' of the environment movement.)

(Adapted from Johnston, Larana and Gusfield 1992: 6–9)

---

way of being politically effective. As with the old social movements, incorporation into a new status quo is likely to be the product of external struggles both within environmental movements and between these movements, business and the state.

General theories of social movements, old and new, should be treated as tools, rather than models into which all experience can be forced. In some situations, they can prove extremely useful, in others, they are inappropriate. There are many specific, diverse and contradictory factors that explain the rise of different environmental movements. There are no overriding, agreed common goals that join all the different movements together. There is no unifying teleological purpose that drives them. There is no single causal reality that made them. Interpretations of the origins and significance of environmental movements are as contested as the movements themselves.

Organisation and structure in environmental movements

Environmental movements vary greatly in both their general and specific objectives as well as in their internal structures and modes of organisation. It is important to consider how these movements hold together internally and the extent to which they coalesce to form more coherent organisations.

When looking at the sprawling activities of environmental movements, political scientists have attempted to use, among other things, pluralist interest group models. But social movements are not just large interest groups or organisations. They are far more complex, are often more diverse, highly informal, amorphous in their structures, and constantly undergoing substantial redefinition (Doyle 1986; Doyle and Kellow 1995). Because interest group models treat collective political action as being driven by shared goals (that is, it is assumed what needs to be analysed), this has inhibited our understanding of the often fascinating mechanics and structures of the more informal relationships that are characteristic of a social movement.

A vast array of informal groups, formal organisations, networks and individuals is involved in each environmental movement. This fragmentation is a reflection of a broad range of differing political ideals and policy goals, and consequent means for achieving them. It also reveals the segmented, diffuse and amorphous nature of the movement's structure. For Pakulski:

> Structures include patterns of links between movement specific groups and organisations, as well as groups and organisations drawn into the orbit of movement activities, but formed independently of them (e.g. political parties, religious bodies, ethnic organisations).
>
> (1991: 32)

For this very reason, it is not possible to ascertain the exact number of environmental groups operating at any one time. The movement is in a

Plate 3 *The Yeppoon Environment Centre (near Shoalwater Bay) on the Central Queensland Coast, Australia. The base from which the battle for Shoalwater Bay was fought. Courtesy of A. Simpson, private collection*

constant state of flux. As issues appear on the political agenda, groups often form. As the issue in question disappears from public view, the group may fade away also. Membership is fluid. Each movement comprises many individuals who are not necessarily 'card-carrying' members of specific environmental organisations. Hence, the overall membership, when defined by different individuals or groups, varies accordingly.

Due to their fragmented nature, a study of the structure of specific movements is an extremely complicated one. There are five different structural forms that dominate movement activity (see Figure 3.1). Each one of these forms has distinct features, and a given environmental movement is made up of the sum of these different structures. The term 'palimpsest' has been utilised to produce a visual representation of this complex set of combinations (Doyle 1991: 3; Doyle and Kellow 1995).[2] Doyle and Kellow write:

> The primary reason for using this word from the Greek is that it has a semantic definition which is useful in the visual presentation of the proposed model. A palimpsest is a parchment from which writing had been imperfectly erased to make room for another text. The net result of

> this practice is a document with several manuscripts still visible, mapped
> unevenly onto each other.
>
> (*ibid.*: 90)

The appropriateness of this analogy seems most apparent in terms of the
structure of environmental movements, with their three-dimensional
space and different levels of political activity found and labelled within
them. Individuals are linked by interconnecting lines on the diagram.
These links depict many, varied networks of individuals. Each network is
different as its definition relies on the perceptions, biases and power
plays of the initiator(s). The key differential variable of the network,
however, is the common goal(s) or ideology that bind(s) the participants
together. In most cases these goals or shared values are specifically issue-
oriented, whether based on an environmental campaign or a particular
type of political system.

These interconnecting lines ignore organisational and group boundaries.
For these networks are fundamentally concerned with relationships
between individuals operating inside and outside other formal and
informal collective forms. Consequently, on occasions, these networks do

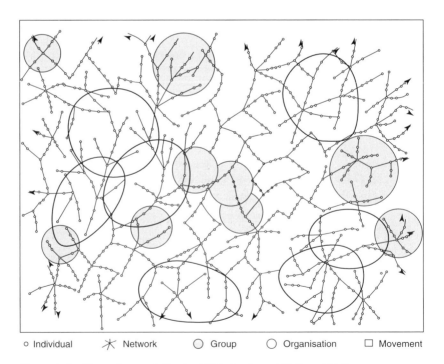

○ Individual　　✸ Network　　○ Group　　○ Organisation　　□ Movement

Figure 3.1 *The palimpsest (Doyle 1991, Doyle and Kellow 1995)*

not intersect with any formal organisational or informal group members. Instead, these links may indicate information exchange and other informal activity. It is these networks that provide the cement which both directly and indirectly links the different parts of the movements together.

Positioned between the levels of the network and the organisation is the *group*. The 'group' shares certain similarities with the network. The group, however, is more permanent than the network; its relationships have become more solidified over time. As a symbol of the more stable status of the group, each group has a perennial title. The creation of a collective, invariant title that symbolises the politics of the individuals and differentiates them from others is a unifying characteristic of groups.[3]

The group also differs from the organisation, which is the fourth level of the palimpsest. It is portrayed on the diagram as a solid, continuous line, symbolising the rigidity and the permanency of its well-defined, often hierarchical structure. These organisations (or NGOs) are the focus of the next chapter and are a component part of social movements. Pivotal to this concept is a vision of a movement existing as some sort of collection of concentric circles, with the organisations (or NGOs) occupying the centre, and all others inhabiting the outer periphery. This is a rather insubstantial perception of what a social movement is, and it illustrates the biases of the organisational sociologist at work. Instead, Piven and Cloward have a far more acceptable, though less defined, differentiation between movements and formal organisations. They state:

> The stress on conscious intentions in these usages reflects a confusion in the literature between the mass movement on the one hand, and the formalised organisations which tend to emerge on the crest of the movement on the other hand – two intertwined but distinct phenomena.
>
> (1979: 5)

The final, and most recently applied, 'manuscript' is at the level of the movement. Each movement is represented on the palimpsest as a square frame. This frame symbolises the boundaries set by each individual. These boundaries are continually changing with time and the different forms of individual, group, organisational and network perceptual definitions.

Environmental movements, therefore, are overarching political forms that include actors advocating all of the diverse eco-philosophies and positions (and many more) touched upon in the previous chapter.

Varieties of environmental movement

Despite the incredible variety of environmental movements operating in countries around the globe, it is still possible to produce a description of the different types of movement and their most common geo-political location and prominence. Such an account cannot be all-inclusive. It is necessary to select the most characteristic forms to use as examples. This means, of course, that there will be a serious degree of simplification and that examples to contradict the generalisation will easily be found. The purpose here is to give a broad-brush account to help locate the character and actions of any particular movement.

Western European environmental movements

Some of the most dramatic, well-publicised and much discussed environmental movements are to be found in Western Europe. There are two major kinds. First, there is the very traditional nature conservation movement. Second, there are the political ecology and anti-nuclear movements, characterised by some writers as part of the 'New Left' (Box 3.3).

There is a major difference between the kind of environmental politics embraced by these two different movements or networks. The nature conservation movement sought 'protection within the existing economic order' (Lowe and Goyder 1983; Rohrschneider 1991: 254). This movement 'proposes reform' by playing politics with a 'green tinge'. The political ecology and anti-nuclear movements demanded systemic change, placing ecological and social objectives above economic concerns.

The nature conservation movement accepted current distribution patterns of both power and economic resources. Both the political ecology and anti-nuclear movements argued for resource conservation along with a more equitable distribution of those resources. The characteristic style of politics of these two groupings was substantially different. Despite the fact that activists within these movements recognise and use, often in mutual criticism, the differences between these political traditions, the public sees both as just part of 'one environmental syndrome' (Rohrschneider 1991: 251–66).

Box 3.3

Major thrusts in Western European movements

Nature conservation movement

The oldest branch of the environmental movement is the nature conservation movement, whose origins significantly preceded the evolution of the Western European ecology and anti-nuclear energy movements. One of the main motives of the nature conservation movement during the nineteenth century was to preserve visible signs of human history in a rapidly modernising world. Other traditional thrusts included bird preservation. More recently, the contemporary nature conservation movement is motivated by pollution problems.

Political ecology movement

This movement emerged at the end of the 1960s in reaction to problems of industrialised societies. There was a strong urban focus, as well as attempts to protect relatively untouched 'natural' areas. By the mid-1970s, environmental organisations usually controlled their operations. As Western European traditional parties were unwilling to incorporate their needs into the governmental process, they formed their own political parties (see Chapter 5). Also in the 1970s, with close ties to other new social movements, it developed the 'new environmental paradigm' view, which included, apart from ecological concerns over economic growth, a critique of science and technology, and a preference for participatory politics. Although there was an interest in the welfare of other species, there was a strong anthropocentric 'human ecological approach', with a 'nation-state' political focus.

Anti-nuclear movement

Although characterised as a New Left movement, in association with the previous category, this movement was primarily born of the desire to shut down power plants in Western Europe (Whyl and Brokdorf in W. Germany and Creys-Malville in France). Also, it attempted to promote the importance of the nuclear energy debate in European politics. It was one of the most radical, unconventional political players in the New Left environmental movement, with an emphasis on unco-ordinated, decentralised actions. On the whole, it did not attempt to play neo-corporatist, formal organisational and partisan politics.

(Adapted from Rohrschneider 1991)

North American environmental movements

Whereas most significant Western European environmental movements have been rooted in anti-nuclear energy issues and the politics of human and political ecology, environmental movements in the United States, Canada, New Zealand, Australia and some parts of Scandinavia have been dominated by wilderness-oriented perspectives with an emphasis on the 'importance of place'. These are not unlike Western European nature conservation movements but with less emphasis on the preservation of the built environment. Whereas the political ecologists of Western Europe have advocated wide-ranging social change along the lines of the four pillars outlined by the West German Greens (participatory democracy, social equity, non-violence and ecology) wilderness-oriented movements have been dominated by a concern with ecology. Part of the reason for this is obvious. These countries still have large tracts of relatively 'undeveloped wilderness' (Eckersley 1990: 70) that can be saved, and a suitably large portion of the population is rich enough to define environmental concern in this way.

What is the environmental movement in the United States like? Obviously, at different times, in different places, the movement has had a variety of different concerns. On the east coast, with such a high population density, it has been impossible to escape from the human element in ecology. Consequently, air- and water-quality issues are rated highly. Unlike the political and human ecology movements, radical, systemic political change is rarely proposed by the movement, dominated as it is by powerful non-governmental organisations (the subject of the next chapter). These NGOs construe nature in a rather limited, instrumental fashion, not unlike the view of government bureaucrats and the corporate think-tanks with which they often work. The 'environmental crisis' is not really seen as a crisis at all but as a challenge for better management. Environmental problems, in this sense, are seen as efficiency optimisation projects played out in the marketplace.

The overriding focus in the western regions of the United States (and this dominates the national agenda) is on wilderness issues. Ecocentric arguments abound in the United States: forest wilderness issues; the reintroduction of mega-fauna (like wolves, grizzly bears and elk); and the 'management' of national parks. With so much emphasis on non-human nature (from both instrumental and intrinsic value positions) there has been little recognition of the fact that humanity is also part of nature.

The mainstream environmental movement in the United States is profoundly apolitical in both the way it perceives and the way it plays environmental politics when compared with the Western European tradition (see Box 3.4).

Box 3.4

The apolitical nature of North American environmentalism

1 Environmentalism is not really seen as inclusive of a human political dimension.

2 There is little criticism of existing political systems.

3 Its pluralist system (see Chapter 1) is so dominant that people do not perceive it as a political model; but rather as *reality*. In this political system, citizens are seen, in their *natural state*, as apolitical. The environmental movement has not been able to escape these dominant cultural perceptions.

4 With such a heavy emphasis on the pluralist model, all citizens are seen as capable of gaining equal access to political power. But to do so, citizens must join a formalised interest group, recognised as 'legitimate' by the state. This leads to a profound reliance on political lobbying.

5 With the heavy emphasis on the apolitical individual, the environmental movement has also fought many battles through the legal system. This has seen the development of a plethora of one-off victories, but it has not seen the emergence of a permanent parliamentary or administrative environmental response. For example, the success of green parties in the United States is virtually non-existent, and there is no cabinet-level portfolio in national government.

6 Another consequence of this apolitical individualism is an attraction to *new age* environmentalism. This form of environmentalism seeks to change values from within the individual. Often these changes are psychological and spiritual. The argument is that individuals must change their inward relationships to nature. While some of these points are salient, unfortunately, they further divorce the movement from the political realm, which requires an interplay *between* social groupings; not just *within* the individual.

7 Finally, where radical change has been voiced (by such organisations as Earth First!) there has been a heavy emphasis on direct, sometimes militant, piecemeal actions.

8 It lacks an international/global dimension. It remains transfixed by domestic environmental issues.

Networks of social ecologists (described in Chapter 2) provide a counterpoint to the more powerful deep ecologists and resource conservationists within the movement, promoting a radical human dimension. The social ecologists of the United States, like the eco-socialists of Europe (though coming from distinct non-Marxist philosophical positions) are raising issues over resource consumption and maldistribution. No longer is conserving nature seen as sufficient. It is now a question of who has access to these conserved resources and on what terms. More recent times have seen the emergence of an environmental justice movement (see Box 3.5). This began with the recognition that *race* was a crucial factor in determining the quality of one's environment. In *Dumping on Dixie*, the ground-breaking work by Robert Bullard, it was argued that black Americans were more likely to live on or alongside toxic waste sites (1993: 32–7). This situation was called environmental injustice and environmental racism. Environmentalism was seen as the exclusive privilege and domain of white people, who were more intent on preserving mega-fauna in wilderness parks than fighting for the basic living needs of the underprivileged black and coloured populations.[4] This part of the movement, more than any other, has dragged other parts of the US environmental movement into a political place where the social ramifications of being green have to be confronted.

Environmental movements of Eastern Europe

The division between Western and Eastern Europe, prior to 'the end of the Cold War', was a key division between liberal democracies with capitalist economies and those countries whose politics was based on authoritarian, state-centred models of socialism.

Before the mid-1980s, there was no environmental movement in Eastern Europe to compare with those in the West. Under authoritarian regimes it is difficult to measure the extent of support for any social movement. There are always movements that remain more or less 'underground' during this period of harsh rule, ready to emerge and flourish when conditions improve. It is interesting to note that it is the non-institutional form of politics that sustains movements under repressive regimes. As discussed in Chapter 1, few NGOs, political parties or other formalised centres of opposition are tolerated in these regimes. Instead, informal networks, groups and associations provide the lifeblood of the social movement's existence. There are no easily defined leaders, no easily located epicentres, and no clear patterns of association to be suppressed. Szabo, in the context of Hungary's greens, writes of this 'structure' of politics:

> The Hungarian ecology movement was born in the mid 1980s, but, given the constraints of mobilizing under the one-party state, it never attained an integrated organization. Rather, it existed in the form of unconnected local citizens' initiatives, single-issue groups, and alternative lifestyle communities. Unlike ecology movements in France and West Germany in the 1970s, there was no unifying antinuclear group in Hungary, despite Chernobyl and the scandals surrounding the only Hungarian atomic power plant in Paks.
>
> (1994: 292)

Major problems in Eastern Europe include air and water pollution, waste management, and soil contamination. These are extremely severe in the 'triangle of pollution': the former GDR (Democratic Republic of Germany, also known as East Germany), the former Czechoslovakia and Poland (Schreiber 1995: 365). These countries are directly responsible for a great deal of the pollution in the North, Baltic and Black Seas. Obviously, this level of contamination poses dramatic health risks. In certain regions of the former Soviet Union such as Lake Baikal, the Aral Sea and industrial centres like Kuzbass, the incidence of cancer 'is up to 50 percent higher than the Soviet average, and respiratory-tract illnesses are common' (*ibid.*: 364). And the levels of environmental degradation and resulting contamination are simply not known, let alone addressed, in the many Asian regions of the former USSR.

Many of these problems are attributed to the *heavy* industrialisation that took place after the Second World War. While Western Europe was involved in developing advanced 'end-of-pipe' technologies, the governments of Eastern Europe ignored environmental costs. Some writers argue that it was the 'command control' decision-making structures of the 40-year-old Stalinist regimes that were responsible for the lack of environmental regulation and reforms (Fagin 1994: 479–94). Whatever the causal factors, severe environmental problems of this kind leave little space for nature conservation or wilderness concerns, although these also exist in the region.

As the control of the Soviet Union collapsed, an environmental movement emerged more openly in Eastern Europe to address pressing problems of environmental degradation and exploitation. This movement

Box 3.5

Principles of environmental justice

Adopted at the First National People of Color Environmental Leadership Summit, 1991, Washington, DC.

Preamble

We, the people of color, gathered together at this multinational People of Color Environmental Leadership Summit, to begin to build a national and international movement of all peoples of color to fight the destruction and taking of our lands and communities, do hereby re-establish our spiritual interdependence to the sacredness of our Mother Earth; to respect and celebrate each of our cultures, languages and beliefs about the natural world and our roles in healing ourselves; to insure environmental justice; to promote economic alternatives which would contribute to the development of environmentally safe livelihoods; and, to secure our political, economic and cultural liberation that has been denied for over 500 years of colonization and oppression, resulting in the poisoning of our communities and land and the genocide of our peoples, do affirm and adopt these Principles of Environmental Justice:

1   *Environmental justice* affirms the sacredness of Mother Earth, ecological unity and the interdependence of all species, and the right to be free from ecological destruction.

2   *Environmental justice* demands that public policy be based on mutual respect and justice for all peoples, free from any form of discrimination or bias.

3   *Environmental justice* mandates the right to ethical, balanced and responsible uses of land and renewable resources in the interest of a sustainable planet for humans and other living things.

4   *Environmental justice* calls for universal protection from industrial by-products and the extraction, production and disposal of toxic/hazardous wastes and poisons that threaten the fundamental right to clean air, land, water, and food.

5   *Environmental justice* affirms the fundamental right to political, economic, cultural and environmental self-determination of all peoples.

6   *Environmental justice* demands the cessation of the production of all toxins, hazardous wastes, and radioactive materials, and that all past and current producers be held strictly accountable to the people for detoxification and the containment at the point of production.

7   *Environmental justice* demands the right to participate as equal partners at every level of decision-making including needs assessment, planning, implementation, enforcement and evaluation.

8 *Environmental justice* affirms the right of all workers to a safe and healthy work environment, without being forced to choose between an unsafe livelihood and unemployment. It also affirms the right of those who work at home to be free from environmental hazards.

9 *Environmental justice* protects the right of victims of environmental injustice to receive full compensation and reparations for damages as well as quality health care.

10 *Environmental justice* considers governmental acts of environmental injustice a violation of international law, the Universal Declaration On Human Rights, and the United Nations Convention on Genocide.

11 *Environmental justice* must recognize a special legal and natural relationship of Native Peoples to the U.S. government through treaties, agreements, compacts, and covenants affirming sovereignty and self-determination.

12 *Environmental justice* affirms the need for urban and rural ecological policies to clean up and rebuild our cities and rural areas in balance with nature, honoring the cultural integrity of all our communities, and providing fair access for all to the full range of resources.

13 *Environmental justice* calls for the strict enforcement of principles of informed consent, and a halt to the testing of experimental reproductive and medical procedures and vaccinations on people of color.

14 *Environmental justice* opposes the destructive operations of multi-national corporations.

15 *Environmental justice* opposes military occupation, repression and exploitation of lands, peoples and cultures, and other life forms.

16 *Environmental justice* calls for the education of present and future generations which emphasizes social and environmental issues, based on our experience and an appreciation of our diverse cultural perspectives.

17 *Environmental justice* requires that we, as individuals, make personal and consumer choices to consume as little of Mother Earth's resources and to produce as little waste as possible; and make the conscious decision to challenge and reprioritize our lifestyles to insure the health of the natural world for present and future generations.

did not emerge simply as a logical response to these problems, serious though they were. Just as important was the fact that environmental discourse was, in part, tolerated by conservative, communist regimes in the initial period, as the 'political dimension to ecology was unforeseen by the regime' (*ibid*.: 480). This initial tolerance of environmental arguments allowed other forms of dissent to manifest themselves. Most accounts of dissent in Eastern Europe treat it as a struggle between the values of the one-party regimes and the 'heroic' forces of democratisation. Szabo supports this point but makes another when he contends:

> On the one hand, social forces have 'unlimited possibilities' for articulating new issues because the inertia of official politics recasts any challenge in terms of the broader drama of democracy versus authoritarianism. In Eastern Europe, the result was the existence of some very limited initiatives that had public and intellectual significance disproportionate to the small number of supporters. These small, powerless groups could become capable of articulating very important – even crucial – but neglected sociopolitical issues. On the other hand, mobilization under authoritarian systems is hampered by the administrative–bureaucratic environment and the use of legal and illegal means of social control.
>
> (1994: 287)

In the early phases, aided by the new tolerance of the 1980s, environmentalists were well received by the broader populus as champions of democracy and dissent. In the case of Hungary, the protest against the building of an Austrian-funded dam on a Hungarian section of the Danube provided a symbolic epicentre not only for the 'ecologists', but for many other forces of 'democracy' (*ibid.*). Environmentalism has been used for many disparate purposes. On this occasion, it was used to undermine an authoritarian communist regime. Interestingly, as the 'velvet' revolutions took place, green forces became accepted as 'legitimate' and were quickly absorbed into the mainstream politics of the new quasi-pluralist systems. Many green interests were accommodated in the new political parties (see Chapter 5). With the demise of state socialism, most people in the East saw more pressing issues than those they narrowly construed as environmental. They were eager to move to the 'free-market' systems of the West, and growth and democratisation were portrayed as the key to the East's success. The arguments of post-industrial theorists ceased to be heard. The environmental movement had served its purpose in the transition period. Chatterjee and Finger observe:

> Not surprisingly, in a highly politicized society and at a highly political moment, the Green movement in the East was first and foremost a political movement with political, i.e. national agendas. Be it in Hungary, Poland, Czechoslovakia, or Estonia, the Green movements turned rapidly into green parties, which in turn quickly acquired a share of national power. But once this had taken place, the Green movement declined. In retrospect, it turns out that the Green movement in the East was instrumental in the transition of the Eastern European countries to a

market economy. Yet despite enormous ecological problems facing the
East, in the 1990s the Green movement has substantially lost momentum.

(1994: 66–7)

But this judgement seems too certain and too early. Some networks
within the green movement have not been co-opted by mainstream
politics and have remained a vibrant though informal voice in Eastern
Europe. This level of non-institutional politics may not be as 'visible' or
as 'reportable' as its more institutional counterpart. It is fluid, amorphous
and flexible. Those concerned by the evidence of environmental
degradation continue their efforts to change the way in which these new
regimes respond to this difficult part of their legacy.

Divisions between movements of North and South

The North–South dichotomy is most simply portrayed as the line
dividing those of the Earth's nations that are wealthy from those that live
in poverty: the haves versus the have-nots. Other nomenclature refers to
the nations of the First World (developed nations) and those of the Third
World (developing nations). Three-quarters of the Earth's population fall
into the latter category. Obviously, this categorisation is oversimplified,
as it is really the North as well as the Southern elites who are wealthy
(*ibid*.). Still, it remains a useful symbolic division. In many nations of the
South, as with parts of Eastern Europe (but, on most occasions, even
more so) the links between environmental degradation, illness and
poverty are obvious. Indeed, basic human poverty (poor nutrition,
inadequate shelter and limited access to education and health services)
remains the most fundamental environmental issue.

Although wilderness-oriented movements do exist, they are largely
overshadowed by debates about the environment/development nexus.[5]
Obviously, the post-materialist thesis is largely meaningless here. There
is little about 'higher values' (in Western terms) when considering the
South's environmental crisis, and crisis it is. Also, little credence is given
to post-industrialist arguments, as most people in the South see the key
problem as lack of ownership of their own resources. Chatterjee and
Finger comment on this point:

> [For] the Third World Network in Malaysia or the Centre for Science and
> the Environment in India, it is no longer industrial development *per se*
> which is considered destructive of the environment. Rather it is the fact
> that development remains controlled by the North instead of the South.

> The weakness of this argument, of course, stems from the fact that it
> mixes together Southern peoples and Southern Elites.
>
> (*ibid.*: 77)

Also, many parts of the South are not heavily industrialised economies,
although this is changing rapidly.

The dominant view of the North is that the poverty of the South has
caused and continues to create environmental degradation.[6] This
environmental degradation is of grave concern to the North, now
advocating global ecology and, as a consequence, seeing itself as
inevitably having to share the Earth's essential survival mechanisms with
the South. Along these lines, the North portrays the major problems of
the South as deforestation, species extinction, global climate change,
desertification and overpopulation.[7]

The unwritten assumption here is that the South is the main
environmental offender, while the North is a model of environmental
controls and reforms. In this view, the North sees itself as bringing its
environmental message (including that of sustainable development and
good management) to the South, to save the latter from itself. With
increased growth and democratisation, 'civil society' will emerge , the
North argues, promoting conditions where people will help themselves.[8]
It is true that Northern environmental activists, mostly through the
vehicles of international NGOs (discussed in the next chapter), are active
in the South. Nonetheless, many successful environmental networks in
the South are inspired by local activists, many of whom would be
construed as 'poor' by Northern standards. They are not degraders or
sustainers, but actors.

In the Philippines, for example, a vast and vibrant local environmental
movement is fighting environmental degradation head-on across a range
of fronts (Broad, 1994: 813). In Africa, grass-roots organisations have
taken a leading role in the struggle against environmental degradation
(Ekins 1992: 114). In India, the movement against the Narmada Dam
(*ibid.*: 88) (which threatened to dispossess entire communities of their
traditional lands) and the Chipko movement were mostly driven by local
activists. The case of the Chipko movement illustrates how a group of
local people, with only the power of their own solidarity, were able to
curtail logging in Uttar Pradesh in April 1973 by hugging trees using the
Gandhian method of *satyagraha* (literally 'firmness in the truth', but
usually translated in the West as non-violent or passive resistance). This
spontaneous movement spread to many parts of the Himalayas over the
next five years (*ibid.*: 143). In South America, there are few better

examples than that of Chico Mendez and his battle to fight for the lifestyle of his community of rubber-tappers against development interests. In Thailand, local people fought the Nam Choan Dam project in the 1980s. Few Western and Japanese investments had been so strongly resisted by the local population since the 1950s. The construction of dams in Thailand had previously led to deforestation, changes in local climatic conditions, declining soil fertility, and degraded water and fishery resources (Hirsch and Lohmann 1989: 445). In Indonesia, the local environmental movement is one of the few dissenting voices allowed in an authoritarian regime that has lasted 30 years under General Suharto (MacAndrews 1994: 369–80). In fact, the list is endless. Poor people are most affected by environmental degradation in the South, and they are the key actors and inspirations of the environmental movement.

Broad writes:

> A reader need only flip through any issue of the British journal *The Ecologist* or the Malaysian *Third World Resurgence* journal to find numerous case-studies of the poor being involved in protecting the environment – replanting trees, struggling against the enclosure of ancestral lands, fighting for indigenous and community resource management.
>
> (1994: 813)

It can be shown time and time again that jeopardising the right of the poor to subsist often leads to environmental activism in the event of most difficult circumstances. Democratisation can create a better climate for environmental action but, as the case of the Ogoni in Nigeria shows, even repressive military regimes face challenges from dedicated and brave poor people seeking to defend their environments.

Although responses to some forms of continued poverty do have a negative impact on local environs,[9] both the degradation and the poverty are generally caused by ancient land-use and ecological histories (such as deforestation, desertification), which are rendered damaging when coupled with the more recent (in human terms) exploitation of the South by the North, and the complex interplay between these factors.

So on the political level, these problems are often the result of many years of imperialism on behalf of the North. The nations of the South were seen as a treasure trove for Northern traders. Now that these nations are 'independent', they have embarked on their own development projects. Over the past 30 years, there have been three distinct periods of development in the environmental movement (see Box 3.6).

Box 3.6

Three stages of development in Southern environmental movements

Stage 1: 1960s

The first development decade in the South brought optimism. Northern-style growth and development were the goals. So much so, that there was little movement opposition within the countries of the South. Movements from outside the South – mainly in the form of large NGOs – occasionally entered the political sphere.

Stage 2: 1970s

During this period, environmental movements emerged in the South. Again, these were dominated by some key NGOs, e.g. the Green Belt Movement in Kenya, the Environment Liaison Centre International, Environment and Development Action in the Third World, and Sahabat Alam Malaysia. Few movement participants opposed the Northern development ideology, but they fought for 'people's development' (another development), not governments' or multinationals' development. This type of development shares many similarities with the political ecology movements in the North, particularly Western Europe.

Stage 3: 1980s to present day

During this period, movements split into two categories. After a period of emphasising local and grass-roots development in the 1970s, many networks in these movements began to collaborate with the government and international agencies again, as in the 1960s. Many coalitions of grass-roots group and local NGOs formed umbrella coalitions e.g. Asia Pacific People's Environmental Network, African NGOs Environmental Network, the Asian NGO Coalition, etc. Many of these powerful coalitions bypass government, on occasions, and negotiate directly with international aid agencies. The other category was the development of environmental protest movements, very similar to the political ecology movements of the North. These networks criticised Northern development schemes. They criticised Northern science and technology, the industrial practices of transnational corporations, national governments, Northern governments, and international aid agencies.

(Adapted from Chatterjee and Finger 1994)

This pursuit of development has been accompanied by environmental degradation for a number of reasons. Damage is partly the result of local industries necessarily producing 'dirty' products in a bid to maintain competitiveness in the 'new global economy', recently deregulated under the General Agreement on Tariffs and Trade (GATT). Multi- and transnational companies, however, are the principal environmental offenders in emerging economies. They can produce commodities far more 'efficiently', using cheap labour and with the less stringent environmental demands of local legislation. These companies also continue to transport their industrial and toxic wastes to these 'developing' nations. Consequently, in 'newly industrialising

Plate 4   *Rare Earth protest, Lismore 1987. Courtesy of W. Brian Alexander*

economies', such as Hong Kong, between 75 and 90 percent of its factories are illegally dumping liquid wastes into the territory's waters (Douglas *et al.* 1994: vii). Between 1.5 and 5.8 million of Hong Kong's residents have experienced health problems due to air pollution. In Bombay, and other parts of India, the air quality produced by industrial wastes is among the worst in the world. Its antiquated water supply system and lack of a proper sewerage system mean that diseases such as typhoid, malaria, asthma and even bubonic plague are rife in the rapidly expanding suburbs, again reaffirming the nexus between environmental degradation and poverty.

Conclusion

There is no single entity that is the environmental movement. There are only many and varied environmental movements, with many and varied networks within and between them. In each country, some movements reflect the dominant cultural and economic aspirations of their national societies, while others project themselves in opposition to these dominant values.

In addition to national and regional distinctions between environmental movements, it is important to understand that since the late 1980s there have also existed global environmental movements that, like the transnational corporations that they either fight or support, no longer have a 'fixed address'. To complete this account of the character of diverse environmental movements, it is necessary to turn to the most formal and visible components, the non-governmental organisations.

Further reading

Chatterjee, P. and Finger, M. (1994) *The Earth Brokers: Power, Politics and World Development*, Routledge, London and New York.

Doyle, T.J. and Kellow, A.J. (1995) *Environmental Politics and Policy Making in Australia*, Macmillan, Melbourne.

Ekins, P. (1992) *A New World Order: Grassroots Movements for Global Change*, Routledge, London.

4 Green non-governmental organisations

- Constitutionalised relationships
- Pluralist and post-modernist theories of NGOs
- Characteristics of green NGOs
- Corporatist environmental policy-making

Introduction

Non-governmental organisations (NGOs) are the most visible players in environmental politics around the globe. They are involved in many different spheres of politics, from the local community level, through the politics of the nation-state, to international politics. They exist in both the predominantly non-institutional domain of social movement politics and in the institutionalised milieu of political parties, administrative systems, governments and beyond.

Although the term non-governmental could include the commercial 'private' sector, the label 'NGO' is rarely applied to business (Bebbington and Thiele 1993: 5). Nonetheless, business may well sponsor sets of NGOs that may make green claims while defending business interests in the policy process.[1]

During the 1970s and 1980s, there was an explosion in NGO numbers. Porter and Welsh Brown estimate that by the early 1980s there were approximately 13,000 environmental NGOs in 'developed countries', 30 percent of which had been formed in the previous decade. Approximately 2,230 NGOs operated in the 'developing countries', 60 percent of which had emerged in the same decade (Porter and Welsh Brown 1991: 56).

Even though these figures are impressive, growth in the NGO sector jumped even more startlingly in the mid to late 1980s. For example, the

African NGOs Environment Network (ANEN) formed in 1982. Twenty-one NGOs were among the founding organisations, but by 1990, membership had swelled to 530 NGOs in 45 countries. In Latin America and the Caribbean, there are currently 6,000 NGOs, most of which have formed since the 1970s. India has some 12,000 'development' NGOs, Bangladesh has 10,000 environmental NGOs and the Philippines has some 18,000 (Princen and Finger 1994: 1–2).[2]

This rapid increase in NGO numbers reflects an explosion in environmental activism. Most NGOs have constitutions, and these symbolic documents give the organisations more permanence than many grass-roots groups and informal networks. Hence NGOs are fairly stable entities and their actions can be more closely tracked and counted throughout a given period. For every NGO there are many more informal groups, associations, coalitions and networks. For example, for the 12,000 Indian NGOs listed above, there are 'hundreds of thousands of local groups' operating in India (Princen and Finger 1994: 2).

Also interesting was the growth in membership numbers within certain individual NGOs during the 1980s. From 1985 to 1990, the membership of Greenpeace (an Amsterdam-based international NGO) increased from 1.4 million to 6.75 million. From 1981 to 1992, Friends of the Earth (FoE) (an international NGO originating in the USA) more than doubled its number of member groups. The Sierra Club (a national US organisation) increased its membership from 346,000 to over half a million between 1983 and 1990. Often these dramatic increases in membership are associated with similar increases in expenditure (*ibid.*: 2–3).

There is some evidence to suggest that there has been a decline in NGO numbers during the 1990s. For example, Greenpeace was substantially criticised for its 'unorthodox' direct marketing techniques and consequently its membership has fallen. A number of other environmental organisations ran into policy difficulties and were exposed to fairly sustained media criticism, often supported by business. The sharp recession of the early 1990s, with increasing unemployment, also put pressure on environmental organisations as economic issues received greater attention. It is also possible that the recent decline in NGO numbers may just be a levelling off after two decades of expansion.

What are environmental NGOs?

The first thing which must be understood is that environmental NGOs are political organisations. An organisation, like a social movement, has distinctive collective properties. The most readily recognisable feature of these organisations is the constitution. Briefly, the constitution is symbolic of both the acceptance of legitimacy and rationality; it publicly states the organisation's intention to work within the dominant structures of the state. Murray Edelman writes:

> On the one hand the constitution legitimizes in morally unquestionable postulates the predatory use of such bargaining weapons as groups possess: the due process of law, freedom of expression, freedom of contract, and so on. On the other hand, it fixes as socially unquestionable fact the primacy of law and of a social order run in accordance with a code that perpetuates popular government and the current consensus on values: the rule of law, the power to regulate commerce, the police power, and so on.
>
> (1964: 18–19)

These constitutions establish certain rules which dictate how power will be dispersed throughout an organisation, how decisions will be made, and how people will relate to each other (see Box 4.1).

Although environmental NGOs are diverse, they are far less heterogeneous than the political groupings spread throughout environment movements, as there are certain shared rules for playing politics. Sometimes, of course, these rules are open to interpretation. The political characters of NGOs are not really indicative of the most radical elements within environmental movements. Many radical activists of a

---

Box 4.1

Characteristics of an organisational constitution

1 It is a very broad statement of intentions.

2 It uses symbolic language to attract a greater number of people, and to diffuse opposition.

3 It can be used to either blunt or focus criticism.

4 It acts as a mechanism of power. It can be used as a tool of repression or as protection against those who attempt to coerce others in the organisation.

more revolutionary bent remain deliberately removed from this more formal realm of politics for two reasons. First, they believe their more radical goals will be co-opted by mainstream politics. Second, they feel better equipped to fulfil their goals within more fluid, less structured political forms.

As with social movement theories, there are numerous explanations of when, why and how organisations form.

Interest groups and lobbying: a pluralist view

The first theory, interest group or pressure group theory, is strongly linked to the pluralist interpretation of liberal democracy. In this view, NGOs are seen as interest or pressure groups. This is rather confusing as we have used the term group, in the previous chapter, to connote a more informal human collective form. But in traditional political science the terms group and organisation have been used interchangeably.

In a pluralist world, it is assumed that there are no significant concentrations of power: power is diffused in society and all citizens have some power resources that can be used to achieve their aims or interests. People will have different aims and ambitions, and different interests and grievances. People with shared visions, grievances or interests are free to form organisations (treated as either 'interest' or 'pressure' groups) to press their claims in the political process. With varying degrees of skill and varying amounts of power, these groups battle for influence. The political system responds to at least some of these demands, and there is a degree of incremental adjustment.

In this model the state is treated as 'neutral', having no particular interests or goals of its own. In pluralist systems NGOs exist to lobby. They do not seek public office directly, but through influence provided by members, public opinion, money and power, they seek change indirectly. Duverger argues that these organisations

> do not participate directly in the acquisition of power or in its exercise; they act to influence power while remaining apart from it; they exert pressure on it. . . . Pressure groups seek to influence [those] who wield power, not to place their own [people] in power, at least not officially.
>
> (1972: 101)

Pluralist accounts capture some aspects of what happens with NGO political action. Many environmental groups do form in response to a shared sense of concern about the harmful environmental consequences of a proposed or existing economic development. Once formed, such groups will use the full array of measures to pressure the political systems to either stop or regulate the damaging economic activity. The problem with the pluralist account lies in its assumption about what happens in the competition with other groups and forces as they struggle to influence government. Not all groups are equal either in their access to power resources or in their access to the political process. Some environmental NGOs do have large sums of money to spend on their campaigns, do employ scientific and legal experts and do have great skill in putting their case using the existing hostile, sceptical media. Greenpeace would be the most obvious and important example, but not many environmental groups are like this. In their contests, these groups are often small, poorer and weaker than their business opponents. In these circumstances, the pluralist framework can produce a misleading interpretation of events. It should also be noted that there are serious problems with pluralist assumptions about the character and role of the state. In environmental disputes, it is very unusual for the state to be a neutral party, seeking to arbitrate between proponents and opponents of a particular project. Quite frequently, government either initiates, fends for or supports a given development, which is then subject to environmental dispute. There is one further point to note about this treatment of NGOs as if they are just pressure groups: to consider environmental NGOs in this limited way takes them out of their movement context and, at best, gives a one-sided account of the dynamics of their actions.

A post-modern theory of NGOs

In general terms, post-modernists see the world as fragmented and in a situation of dislocation and crisis. Just as the post-materialist theory of environmental movements comfortably sits alongside pluralist perceptions of politics, the theories of post-industrialism are more easily accommodated into a vision of a post-modern polity.

There is now an intense fragmentation of political processes and actors, as evidenced by the description of the palimpsest presented in the previous chapter. Let us briefly look at the post-modernist theory of NGOs posited by Thomas Princen and Matthias Finger. Unlike the pluralists, Princen and Finger believe that there is an environmental crisis, which demands a new form of politics as a response. As well as all the standard economic and technological changes that transform politics, Princen and Finger note:

> In addition, the global ecological crisis has reinforced and accelerated this process toward post-modernism, rather than reversed it or slowed it down. It has led to more fragmentation, further eroded collective projects, and contributed to the multiplication of social environmental actors.
>
> (1994: 61)

They argue that the project of modernity is under attack and there is an 'absence of a common reference point'. Nation-states, particularly within any globalised economy and facing global ecological problems, have lost the capacity to be effective in solving environmental problems. More than that, nation-states are often dangerous to the Earth's ecology. Their primary functions are twofold: first, to provide a basis for an expanding military–industrial complex; second, to promote the notion of a national economy, based on 'growth' and 'progress' at all costs. Princen and Finger argue that the search for effective and substantial environmental reform has to be pursued in a domain beyond and below the nation-state level. Their predominant answer is to value the style of NGO politics that has emerged from this fragmented and diffused political situation.

> As I suggest that environmental NGOs free themselves from traditional politics, change the reference point and privileged means of action, grow in numbers and inter-connectedness, and become increasingly transnational, they contribute to societal change and transformation in another way: they become agents of social learning. . . . Indeed, rather than focusing on traditional politics, how to influence it and how to mobilize for it, environmental NGOs build communities, set examples, and increasingly substitute traditional political action. They become agents of social learning, whereas social movements were actors of political change only.
>
> (*ibid.*: 65)

While much of what they write evokes the non-institutional experience of social movement politics, Princen and Finger see NGOs as largely separate from social movements. They contend that it is not only the traditional defenders of the modernist project but also the social movement theorists who remain bewitched by the nation-state model of politics. They are right about this point, and many NGO theorists remain equally transfixed by pluralist and corporatist theories of access to the politics of the nation-state. The model of social movements discussed in Chapter 3 is not confined to what goes on inside nation-states but does include NGOs in the environmental movement. The real advantage of Princen's and Finger's account of separate and distinct environmental NGOs is found in its contrast with pluralist assessments. Whereas the pluralist model emphasises lobbying for incremental change, post-modernist theories, like Princen's and Finger's, see NGOs as possessing far more direct, creative and transformative powers. Further, post-modernists see social movements and environmental NGOs forming as a result of fundamental ruptures within the modernist project, whereas the pluralist 'interest group' theorists regard them as a minor variation on the politics-as-usual theme.

Depending on the models of power used to explain the emergence and definition of NGOs, there are different answers to key questions about green NGO politics. Each model presupposes very different notions about what NGOs actually do and what tasks they usually perform, and each promotes a clearly separate view of the relationships between NGOs and the state. Finally, each framework gives a different perspective on the internal dynamics of NGOs. Before directly addressing these topics it is useful to develop several broad criteria that can provide a preliminary typology of green NGOs practising all around the globe.

Typology of green NGOs

Although not as diverse as environmental movements (as they share common organisational characteristics), environmental NGOs vary on the basis of an array of determining factors. These include:

1 their geopolitical origins;

2 their political ideology;

3 their size;

4 the level of their political focus;

5 their funding sources;

6 what they provide (what tasks they actually perform);

7 their internal politics/structure; and

8 their relationships to the state.

Let us briefly review the first five characteristics with reference to Tables 4.1 and 4.2.

Table 4.1 Typology of green NGOs in the North

NGOs	Greenpeace	FoE	ACF	Jatan	Earth First!
Characteristics					
Geopolitics	North, North in South	North, North in South	North	North, North in South	North
Internal structure/politics	Hierarchic, top-down	Dispersed confederation of regions	Federation of state branches and chapters	Centralised	De-centralised
Political ideology	Radical political ecology	Radical political/social ecology	Reformist, resource conservation	Reformist, resource conservation	Radical deep ecology
Size	5 million members	50 member groups	100,000 members	4 staff, vast network	cannot measure
Political level	Transnational	Transnational	National	International/National	Regional
What they do	Non-violent direct action, mass mobilisation	Mass mobilisation	Lobby appeal to elite	Mass mobilisation	Sometimes violent direct action
Relationship to nation-state	Few	Few	Close	Few	None
Funding source	Membership drives, direct marketing	Membership, private	Membership, government, direct marketing	Network funding	Newsletter, low costs

Table 4.2 *Typology of green NGOs in Eastern Europe and the South*

NGO Characteristics	*EFT*	*Green Don*	*Grameen*	*BRAC*	*DAWN*
Geopolitics	East West in East	East	South	South, North in South	South
Political level	National	Local, regional	Regional, national	National, international	Regional, national
What they do	Conferences, scientific research	Public education	Providing env./dev. funding	Env. and dev., widespread direct provision	Women's community env. issues
Funding source	70% int. World Bank	None	Investors	International, government	International community

Geopolitical origins

Obviously this denotes the geopolitical region in which the organisation either formed or operates. The 'place of origin' is important, as many large organisations, particularly those involved in global ecological issues, increasingly see themselves as transnational. Some organisations, although originating in the North, are major operators in the South. In more recent times, Southern NGOs have also established extensive global networks.

Political ideology

With reference to Table 4.1, there is a basic difference between radical and reformist organisations. Also, under these broad headings there are a myriad of ecophilosophies: from political ecology to resource conservation; from deep ecology and ecofeminism to social ecology. Obviously, in different parts of the world certain NGO ideologies will be more apparent.

Size

Green NGOs vary dramatically in size. For example, Green Don (Zedon) (Table 4.2), a Russian NGO operating in the Black Sea, has 40 members and no operating budget (Global Env. Facility Black Sea Env.

Programme 1995), whereas Greenpeace (Table 4.1), with over five million members, has had an annual budget of approximately 100 million US dollars.

Political level

Some organisations are involved in policy making at a global level; some aim at national politics; some are regional and some are intensely local, forming to resolve, for example, an environmental issue indigenous to a tiny community catchment area.

Funding

Many NGOs generate their own funding through membership dues. Others are more aggressive in the marketplace, soliciting money directly from the public in the form of donations, bequests and merchandise sales. These direct marketing techniques are more prominent in the North. NGOs, in both the North and the South, also receive some money from their national governments. In the South, many NGOs receive money from Northern NGOs and aid agencies, often bypassing their national governments. This sometimes creates tensions when certain NGOs are seen by governments as 'workers for foreign interests'.

For the remainder of the chapter, let us consider in more detail the final three differential variables: what do green NGOs actually do; what are their internal politics and what is their relationship to the state?

What green NGOs do: North and South

Neo-pluralist models are far more appropriate in describing the role of NGOs in the North than in the South. NGOs lobby in numerous forums. Sometimes they lobby local or national politicians directly, by using their potential electoral powers. On other occasions, NGOs lobby administrators in government departments, who are often more 'permanent' than their elected counterparts. At times, particularly in the world of international affairs, lobbyists work at influencing diplomats or scientists. Others try to influence the policies of political parties and corporations or try to convince a certain community that one environmental action is better than another. Finally, many NGOs lobby other NGOs, trying to get some agreement on green objectives and strategies.

Brian Martin refers to this emphasis on lobbying as 'the appeal-to-elites' method (Martin 1984: 110–18). In appealing to elites, environmentalists have to speak a similar political language to those already inhabiting the halls of power. Table 4.1 refers to the Australian Conservation Foundation (ACF), probably the most important mainstream green NGO in Australia. During the years of Labor Party rule (1983–1996), the ACF worked very closely with government and it was widely seen as politically influential and politically aligned in a party sense. NGOs working in a close relationship with government necessarily define their environmental goals in terms that make sense to the status quo and, in turn, the status quo operators can gain entry into the environmental movement.

Although pressure group politics is the dominant form of politics played by Green NGOs in the North, it would be wrong to assume that it is the only form of politics. In the European political ecology movement, there are many actions that are aimed at mass mobilisation rather than appealing to elites. Sometimes mass mobilisation is used as an advanced form of lobbying. On other occasions, mass mobilisation strategies are based on the assumption that more widespread change is needed than is assumed under the pluralist model. These changes include alterations in mass consciousness and individual value systems. As a result, some of the more radical NGOs in the North involve themselves in mass education programmes.

The direct actions of Greenpeace (Table 4.1), augmented by shrewd utilisation of mass media, were designed to attract widespread support from the global citizenry, rather than gaining favour directly with elites. Earth First!, a US direct action NGO steeped in the ideology of deep ecology, is similarly engaged in this form of protest (Table 4.1).[3] Unlike its Northern counterparts, however, it does not rule out militant actions (most specifically property damage) in its efforts to 'protect Mother Earth'. This method of eco-sabotage is almost uniquely an American phenomenon. Most direct-action NGOs remain committed to the principles of active non-violent resistance (Doyle 1994).[4]

Obviously, this more radical, mass-mobilisation role of NGOs is more readily explained by the post-modernist model of politics. In this time of fragmentation and dislocation, some NGOs are seen as deliberately bypassing governments and acting more directly at local, regional and global levels. But acceptance of this post-modernist model cannot be automatically equated with more radical forms of politics. Sometimes, by rejecting government links, NGOs are also entering the more socially conservative world of the marketplace. The appeal-to-elites method of lobbying governments is replaced by appeals to corporate elites. On some occasions, there may be little to separate the demands of business and the goals of the state, but it is still useful to distinguish between them. For example, instead of lobbying governments with a hope that they will monitor and control more closely the deeds of large corporations (through legislation), some NGOs in the North are dealing directly with these corporations. Chatterjee and Finger discuss this increasingly close relationship between big business and some of the less radical Northern NGOs:

> WWF, for example, received $50,000 each from oil companies Chevron and Exxon in 1991. The National Wildlife Federation conducts enviro-seminars for corporate executives from such chemical giants as Du Pont and Monsanto for a US$10,000 membership fee in their Corporate Conservation Council Programme. The Audubon Society meanwhile sold Mobil Oil the rights to drill for oil under its Baker bird sanctuary in Michigan, garnering US$400,000 a year from this venture.
>
> (1994: 70)

These more free-market interpretations of environmental reform are culminating in some NGOs no longer seeing themselves as non-profit organisations but as 'players' who can trade freely in the marketplace (see Box 4.2).

With increased income derived from deals, investments and sales of organisational products, some Northern NGOs are privately buying tracts of land for wilderness preservation and other nature conservation purposes.[5] Also, Northern NGOs are directly funding governments and NGOs in other countries, most particularly those in the South (but also in the East). One example is the way in which US NGOs are buying the debts of poor nations in direct exchange for protected wilderness areas. Schreiber explains this process:

> US environmental groups have bought approximately $1 million of Bolivian debt for $100,000. This has now been offered to Bolivia in exchange for an enlarged nature protection area and a bill for stricter environmental protection in such areas. Similar (debt-for-nature) agreements have been made with Costa Rica, Ecuador and the

Philippines. The first debt-for-nature swap with a formerly socialist country, Poland, is now underway. The World Wildlife Fund (WWF) plans to buy US$1 million of Polish debt for $100,000.

(1995: 378)

Box 4.2

Twelve richest national environmental organisations in the USA

Ron Arnold, leader of the 'Wise Use Movement' (a long-time critic of the environment movement) released a book entitled *Getting Rich*, an analysis of the environmental NGOs' income, salaries, contributors and investment patterns. Sometimes, these organisations are called the 'Big Ten' or the 'Dirty Dozen'.

The average compensation for top executives was US$174,873 in 1994–95. These twelve organisations have assets of over 1.2 billion dollars. The report lists the major donors to these NGOs. There are many Fortune 500 firms: ARCO, Boeing, Dow Chemical, Du Pont, Exxon, Shell Oil, etc.

Justin Time writes: 'For their part, the groups return the favour and hold major investments in the aforementioned behemoths. TWS (The Wilderness Society) for example, has $385,000 in GMAC and major holdings in dozer-builders Deere and Co. and Cummins Engine, JP Morgan, 3M, US Bancorp, Southern Cal Edison, GE, Gannett and IBM.'

'British Columbia's favourite deforestation company Macmillan-Bloedel recently bought 50 per cent of Truss-Joist. Truss-Joist's CEO, Walt Minnick, put up $250,000 for a seat he now occupies on the TWS Board of Directors.'

(Adapted from Justin Time and the Mystery Riders Workshop 1995: 10)

Despite the more recent emergence of forms of environmental politics that could be explained by a post-modernist view, NGOs in the North remain primarily a political lobby, still identified as pressure and interest groups, and dominated by a pluralist 'way of seeing'.

In other parts of the world, green NGOs are involved in very different forms of eco-activity. Of course, there are still the lobbyists and the mass educators, but in many nations of the South, NGOs are the direct providers of infrastructure. For example, in many cities in Asia (like Bandung in Indonesia, Bombay in India, or Bangkok in Thailand) where there is a massive population explosion, entire cities are forming outside of the established 'city limits' (Douglass, 1994: ix–x). In these cases, NGOs are directly involved in the provision of clean water; the physical

Plate 5 *Protesters approach Roxby Downs uranium mine, South Australia 1984. Courtesy of Kevin Wood/Developments*

labour of cleaning up refuse and the disposal of solid wastes; the building of shelter and the provision of sewerage systems; treating people directly for disease and malnutrition; direct provision of food and other basic essentials for living; and co-ordinating many other 'hands-on' tasks and activities.

Some Southern NGOs act as conduits for the purposes of larger Northern NGOs. It is interesting that Northern NGOs increasingly accept the importance of 'community-driven projects' in other countries but do not feel this important in their own countries. The funding received by Southern NGOs directly from the North has been quite large in the past, although it has been decreasing in more recent times as the South's 'independence' leads it to find new ways of funding its own 'development' projects, coupled with the 'turning off' of funds from the North.

With poverty seen as both a cause and an outcome of environmental degradation, many direct programmes are initiated by Southern NGOs to create an awareness of social advancement, economic emancipation and income generation for the poor.[6] One way that enables poor people to resolve some of their community problems relates to the availability of credit. One NGO operating in Bangladesh, the Grameen Bank (Table

4.2), is an excellent and successful example of extending banking facilities to the poor. Ekins describes this NGO as follows:

> The Grameen Bank was started in 1976 by . . . Muhammad Yunus who wanted to prove that . . . the poor were an eminently bankable social group. . . . Grameen specifically targets the landless poor and especially the women among these: about 90 per cent of its borrowers are female. . . . By September 1990 there were 754 banks in over 18,500 villages. . . . Up to September 1990, 85,500 houses had been constructed with these loans, and US$22.7 million lent.
>
> (1992: 122–3)

Just as the key environmental issues are perceived differently in the South, so too are the roles green NGOs play. In many places, governments simply have not been able to respond to the environmental problems in their countries. NGOs, both local and transnational, have filled this void. Consequently, the neo-pluralist and corporatist issues of 'access to the state' through lobbying are largely redundant in the discussion of environmental policy. Instead, the model of a post-modernist polity provides a set of analytical tools that appears far more appropriate in explaining what NGOs actually do in the South: they de-emphasise the role of the state; they directly target local communities; and they build transnational networks with Northern and other Southern NGOs. Princen and Finger write of the less institutional focus of NGOs operating in the South:

> Just as Northern NGOs are becoming more institutionalized, Southern NGOs are building organizational skills and financial independence and, as a result, increasingly demand greater autonomy and less dependence on Northern supporters. . . . As Southern NGOs are becoming more independent and setting the international agenda, Northern NGOs are looking to the South for ideas, as well as to establish their own international credibility.
>
> (1994: 8)

Internal politics and structures

There has been much study by political scientists of the relationship between NGOs and the state, but there have been few studies of the internal political dynamics of NGOs. Often internal politics determines the external politics of an organisation: whether it pursues a close relationship with the state and/or certain businesses, or whether it maintains its independence.

To the pluralists, interest groups form to create incremental changes within a political framework dominated by the state. Most are seen as single-issue organisations, not pushing an agenda of systemic political change, as such. As a result, the internal structures of NGOs seeking to lobby governments will often be very similar to governments themselves. Internal structures, both formal and informal, reveal how decisions are made within organisations and how power is distributed. For example, NGOs in the pluralist picture will usually have 'top-down' decision-making structures, with clearly defined leaders and formalised procedures; their administration will be similarly centralised and bureaucratised (with clear differentiation between tasks), and there will be a clear separation between members, elected honorary officials and professional bureaucrats. On occasions, NGOs are so closely intertwined with the state that they base their offices in the same cities where governments centralise their administrations and may mimic national electoral procedures in their own organisations.

Those NGOs whose activities could be explained using post-modernist models of power would remain distant from the state for two reasons. First, they would see the level of nation-state politics as increasingly meaningless in terms of resolving complex global ecological problems. Second, they are crisis-oriented (and often revolutionary by nature) and would contend that playing politics in a mainstream fashion could only lead to incremental changes, which would not be sufficient to solve complex environmental problems.

Obviously, if one were to construct environmental politics in post-modernist terms, one would demand far more from the internal workings of green NGOs. As Princen and Finger would argue, green NGOs have to provide an example to the rest of the world, and this includes how they empower their own people. So rather than reflecting the mainstream political structures (and forging a closer relationship with the nation-state), their internal dynamics must simultaneously challenge status quo politics and provide new political pathways.

Internal politics are most visible when particular cases are explored. Two examples of the pluralist pressure group style NGOs are the Sierra Club in the United States and the Australian Conservation Foundation (Figure 4.1). Both organisations work closely with governments and, at times, have been heavily funded by them. Both see themselves as fundamentally lobby groups and educational organisations and their structures reflect this mainstream politics. Both organisations are even structured along federal lines, reflecting the state electoral divisions

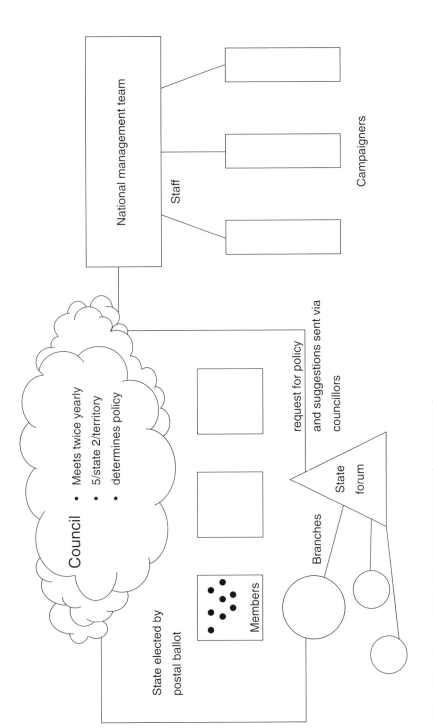

Figure 4.1 *The Australian Conservation Foundation (structure)*

found within the USA and Australia. Both organisations, although seemingly directed by officials elected by the membership, are dominated by career administrative professionals.

Friends of the Earth's (FoE's) decision-making structure (Figure 4.2) reflects its political ecology background.[7] FoE seeks to challenge advanced industrial societal ways of living and, consequently, has a more post-modernist perspective on its role as an NGO. As part of this, it promotes participatory democracy. Indeed, participatory democracy is a central plank in many green NGOs. FoE does not possess a huge number of members but seeks to involve its members, as much as possible, in the everyday running of the organisation.[8] Its administration attempts merely to provide co-ordination rather than a centralised determination of policy. The loose confederation of regional groups is given a large amount of autonomy to set their own political agendas. Although FoE is an organisation, it is almost as non-institutional (at least in theory) as an informal group or network. This informality and lack of centralised control reflects a radicalism in its green goals. There is a strong sentiment in these more radical NGOs that if environmental issues are to be genuinely resolved, then these political changes begin at the intra-organisational level.

It would be simplistic to suggest that all NGOs not pursuing environmental changes in partnership with the state are more radical in their internal political form. Greenpeace, for example, is also an organisation which believes that environmental problems cannot be resolved by a direct appeal to government elites. Unlike the political and social ecological organisations, however, it has not traditionally advocated systemic changes. Rather, as evidenced by its direct-action campaigns, it relies heavily on the existing media to broadcast its often issue-specific, mass-mobilisation campaigns. But by viewing its structure (Figure 4.3) one can immediately see that the organisation is highly centralised, with very little room for dissent among its operatives. Like the US 'Big Ten', it models itself less on government (or open democratic structures, like FoE) and looks remarkably like a large corporation. Its members are basically magazine subscribers with almost non-existent powers for influencing organisational policies. Greenpeace responds to these charges from other environmentalists with the hard to dispute fact that 'it gets the job done'.

Whatever the differences in these pluralist and post-modernist models, most NGOs experience tensions in occupying two worlds: the non-institutional realm of social movements and the more institutional

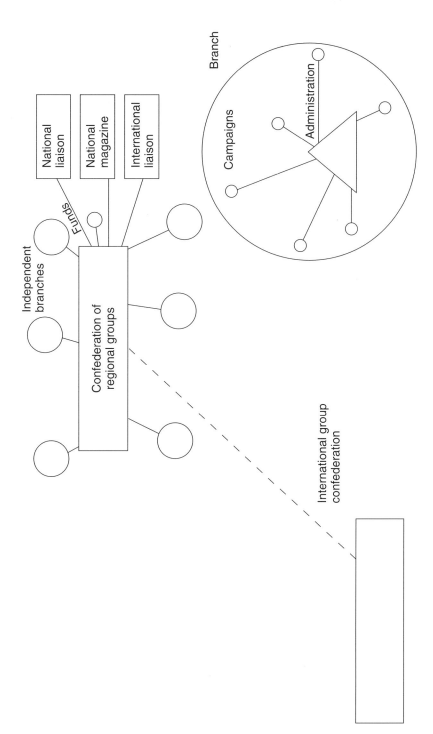

Figure 4.2 *Friends of the Earth (structure)*

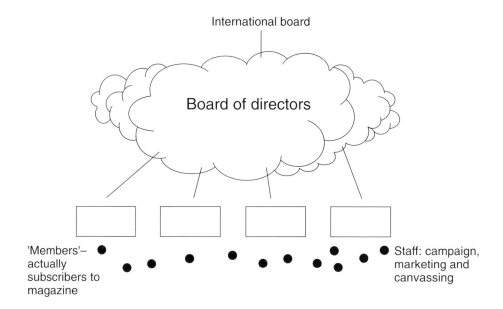

Figure 4.3 *Greenpeace (structure)*

world of governments. In organisations like the Sierra Club and the ACF, which hold intimate connections with, respectively, the United States Democratic Party and the Australian Labor Party, there is an ongoing debate about being too close to political parties, governments and corporations. Additionally, there are often conflicts over levels of professional control versus the right of members to direct their own organisation. The case of the Sierra Club is most apt here. In recent times, a breakaway group has emerged from within the Sierra Club called John Muir's Sierrans. This organisation was formed to 'reclaim the Sierra Club and restore Muir's (the founder of the Sierra Club) vision, passion and strength' (Hansen 1995: 26). Chad Hansen, a key mover in the society, writes about the Sierra Club management:

> At some point they lost sight of the fact that ecosystems are more important than the organization; that this is not about their organization, but a struggle with life and death consequences for millions of species and future generations of humans. They succumbed to the intoxications of political access – at a price. They lost their vision. They lost their connections to the land. After a while, there ceased to be a right and wrong – just access, ego, and power. While there's still hope – and there is still hope – it's time for this to change.

> (*ibid.*: 26)

Two points are raised by Hansen. First of all, green NGOs are rarely unified. Obviously, less conflict is tolerated in the more highly structured organisations than in informal networks but internal political and ideological conflict still remains regardless of the efforts to quell dissent. Even within an organisation like Earth First!, which is internationally recognised for its direct-action tactics, there are competing ideological factions, the Wilders and the Holies. Taylor explains:

> Wilders . . . focus exclusively on wilderness, and thereby, in their minds, on biodiversity and biocentrism. . . . They consider themselves true patriots, trying to preserve the sacred landscape of America (and don't rule out militant forms of protest). . . . Opposite of the Wilders . . . the Holies . . . insist that a 'holistic' perspective is needed; one has to examine how threats to biodiversity are related to other social issues (they rule out militant strategies).
>
> (1991: 263)

The second point derived from Hansen's comments relates to the pressures to formalise. Elite theorists like Robert Michels would say that oligarchy (rule by the powerful few) is the inevitable result of any organisation. He referred to it as the 'iron law of oligarchy': 'whatever says organisation, says oligarchy' (Michels quoted in Plamenatz 1973: 32). There can be no doubt that whether an NGO has grass-roots origins or was 'born institutionalised', there are strong external pressures to mimic the politics of the status quo. But for every Sierra Club or ACF, there are also examples of organisations like Friends of the Earth that continue to promote the concept of direct democracy, despite pressures to the contrary. Rather than iron law, there is a strong tendency towards oligarchy.[9] Obviously, in those organisations that reproduce the political forms of governments and/or corporations, these pressures are greatest. It is for this very reason that many of the more radical environmental groups and networks deliberately remain non-institutionalised. Formal organisation brings advantages and disadvantages to environmental politics. Increased short-term effectiveness and efficiency often comes with the loss of long-term vision, and power becomes concentrated in the hands of a few.

Relations with the state

So far, this chapter has concentrated on the basic and internal organisational characteristics of environmental NGOs. Now it is necessary to consider the characteristic relationship between those NGOs and the state. Pluralist models of power and politics describe a world in which what the state does is a product of the pressures brought to bear on it by various competing groups and social forces. These groups put pressure on the state but form no part of the state system. In the 1970s and 1980s, such accounts were challenged by the rise of various kinds of corporatist analysis. Initially, corporatist theory simplified the assumptions of the pluralist world. Instead of a multiplicity of competing groups, emphasis was placed on just a few, basically those of business and labour. Attention was then paid to the interactions between business, labour and the state in various kinds of corporatist situation. From this flourished a whole literature, extending the corporatist/incorporatist model to a whole range of different policy issues, including environmental organisations and environmental issues.

In general terms, there are two contrasting evaluations of this change. Some corporatist theorists see this situation as most positive for 'civil society' as the process 'entails the state's recognition of the functional interest groups and their incorporation into the process of policy making and implementation' (Capling and Galligan 1991: 57). Others are concerned that NGOs are threatened by the process and interpret it as an effort by the state to incorporate and neutralise opposition.

In a study of Australian environmental policy over a decade, McEachern identified three different ingredients (see Box 4.3) that make up this corporatist process: incorporation, assimilation and adaptation.

McEachern's ingredients, although derived from the Australian context, would seem to apply to the ways in which environmental policies have been shaped during the 1980s and 1990s in a variety of liberal democracies. Even at the international level, in those forums sponsored by the United Nations, this description of the corporatist policy process seems quite apt in explaining the relationship between green NGOs and nation-states and their shared efforts in environmental policy making.

For example, Chatterjee and Finger, in *The Earth Brokers*, describe a comparable policy-making scenario at the United Nations Conference on Environment and Development held in Rio de Janeiro in 1992. Many NGOs were invited to the conference. Accordingly, many observers equated this with more power being given to the NGOs by the participating nation-states. Some were admitted into the process of formulating the policy documents, while others participated in an alternative NGO gathering in the same city. Despite the environmental

---

Box 4.3

Three ingredients in corporatist environmental policy making

Incorporation

Environmental activists and the business community are brought together inside a set of 'normal' political negotiations defined by the nation-state or a combination of states. These negotiations establish the sense that there is a consensus position on environmental concern. In the act of negotiation, dissident sections of the environment movement and business (that is those who do not believe consensus is possible and/or desirable) are constructed as just that, dissidents against an emerging, shared, politically acceptable position.

Assimilation

Second, there is the process in which the socially critical discourses of ecology and environmental concern are taken and turned into legitimate, acceptable, non-threatening discussions about existing economic and resource development practices. The initial formulation of 'sustainable development' was of this kind.

Adaptation

This ingredient necessarily complemented assimilation in that it involved a consideration of the evidence of environmental damage drawn from the arguments of environmental concern. This evidence of damage could then be assessed by relevant governments and scientists and incorporated into policy considerations. Although it would have been possible for politicians to marginalise environmental concern, a potentially more effective response was to accept the evidence of damage and to seek policies that would allow maximum economic growth while minimising damage.

(Adapted from McEachern 1993: 180–1)

NGOs being allowed to take part in the debates over the wording of *Agenda 21*, Chatterjee and Finger believe that the process of corporatism weakened their political capacities. They were now seen as endorsing the policies of UNCED. In turn, this could be used to legitimate the activities of big business and Northern governments. Opposition and dissent were marginalised, extracted and discarded from the policy debate, by selling the notion that the economic imperative of growth and development (sustainable development) would actually enhance the environment. Chatterjee and Finger observed:

> This, in our view, was the result of a long-term transformation of the Green movement world-wide, combined with the very way the UNCED process was set up, as a means of reducing potential protest by feeding people into the Green machine . . . if there was no substantive outcome in terms of conventions and documents, UNCED was at least an exercise in mobilisation and cooptation, weakening the Green movement on the one hand while identifying and promoting potential opponents – mainly from the South – on the other.
>
> (1994: 100, 103)

The more general debate over corporatism and incorporation suggests the limits of these arrangements for environmental NGOs. On the one hand, moving into such corporatist forums gives an NGO access to information, funds and significant decision makers. Here is a chance to be influential and have some input into the content of important environmental decisions. On the other hand, there is the danger of incorporation. The symbol of an environmental NGO's co-operation may be of far greater value to government than any specific decision made. In some cases, the ability of an NGO to respond to environmental problems may be impaired by taking part in a corporatist project. Such participation has costs in terms of time, resources and energy that cannot then be spent on direct mobilisation.

Conclusion

To summarise, there has been an explosion in the number of NGOs dedicated to environmental concerns in the last two decades. NGOs are organisations and, as such, share a collective political form with other organisations, based on the existence of a constitutional charter. Despite this shared reality, they are still quite diverse in their characteristics: their geopolitical origins; their ideology; their size; the sphere of their political activity; their funding sources; and, most significantly, they differ in what tasks they actually perform. Finally, their internal structures (often informed by their goals and ideology) dictate the extent to which they establish relationships with governments and/or business corporations.

In recent times, NGOs have taken on an increasingly important role. With the emergence of global ecology, many environmental issues are seen as beyond the traditional scope of national governments. There have been attempts by governments to respond to global environmental problems through ventures like the United Nations' Rio Earth Summit and subsequent conferences on human populations and habitats. But governments are still lagging behind in their responses, and this transnational political space has been occupied by corporations and NGOs, which can cross nation-state boundaries more readily. This globalisation of ecological and market systems has led to 'the politics of no fixed address' (Doyle 1996).

Whereas traditional interest groups sought to influence those in government indirectly, green activists, some of whom have operated within NGOs, have decided to seek more direct entry into established forms of institutional power, including the formation of green political parties and taking part in elections.

Further reading

Lowe, P. and Goyder, J. (1983) *Environmental Groups in Politics*, Allen & Unwin, London and Boston.

Princen, T. and Finger, M. (1994) *Environmental NGOs in World Politics: Linking the Local to the Global*, Routledge, London and New York.

5 Political parties and the environment

- ○ **Electoral systems and politics**
- ○ **Political parties**
- ○ **Green parties**
- ○ **Electoral success and failure**

Introduction

> Political parties provide the primary method of selecting political elites
> and largely determine the content of electoral and legislative agendas; in
> addition, the parliamentary structure of most European governments
> converts partisan majorities into control of the basic institutions of
> governance.
>
> (Dalton 1994: 213)

Russell Dalton's comments are on events in Western Europe, where the
most successful green political parties are to be found.[1] Political parties
are central to the working of Western liberal democratic representative
systems. The act of governing is driven by the actions or demands of the
major political parties. Government in these systems is largely the result
of party confrontation: 'It is party confrontation – automatic, blind,
occasionally senseless confrontation, regardless of the nature, purpose or
importance of the issue' (Jaensch 1983: 215).

There are three basic ways in which environmentalists have responded to
the electoral system: (1) by consciously abstaining from electoral politics;
(2) by influencing existing political parties to take on elements of their
ideological package (portrayed here as playing normal politics); and (3)
by creating green parties. This chapter considers each of these positions
in turn and concludes with an assessment of the relative costs and benefits
of each of these choices.

Rejecting electoral politics

Many environmentalists believe that formal, traditional politics is incapable of resolving ecological problems and partisan party politics exemplifies traditional politics. Party politicians are often perceived as having little interest in resolving difficult environmental problems or pursuing principled activities. Instead, they are seen as self-interested power seekers; strivers for power in a bid to create more power, as a means of staying in power. The best way of attaining this power, this argument continues, is by providing a party platform that offends as few people as possible and certainly not those who hold significant power in society. Often major parties aspiring to win government begin to look more and more alike. Mayer suggests that this process is to be expected:

> It is not surprising that the two parties which make up the 'ins' and 'outs' should come to resemble each other markedly. Each, in striving for electoral support, tries to antagonise as few voters as possible and to win the support of as many as possible, and hence is driven to become moderate.
>
> (1969: 19)

If this is to be expected, then it should not be surprising that established political parties fail to challenge the status quo when it comes to environmental problems and that such parties make little effort to define and achieve green political goals.

Most environmentalists realise that there is little chance that greens will gain government in the foreseeable future. So why play a political game that one cannot win? Even more apposite is the argument that ecological concerns are universal and must be accepted by all sides of politics. There is a fear that supporting one party will inevitably lead to a backlash, where the 'wronged' party will become hostile to environmental concern and good environmental practice. In addition, there is concern that not only the goals but also the manner of playing traditional party politics in traditional systems is adverse to the new politics of environmentalism.

In an article strongly urging environmentalists to refrain from playing politics in electoral terms, Brian Martin identifies seven points, which form the content of Box 5.1.[2] For these reasons and others, some French environmentalists argue: 'Elections: piege a cons' [Elections: trap for idiots] (Dalton 1994: 220).

Plate 6 *Activist locked on to woodchip conveyor belt, Midways export facility.*
Courtesy of Jock Strauss

It is interesting that Martin's position is largely based on Australian
social movement experiences. Although involved in an active and, at
times, successful movement, many Australian environmentalists have
been extremely reticent about forming their own green party for many of
the reasons Martin espouses. Coupled with this reticence, there is a
strong belief in the potential of the dispersed, connected, informal and
longer-term politics of the social movement collective form. It was not
until 1992 that a national green party was formed, and when compared
with European green parties, it remains in its infancy.

Sometimes, the decision to abstain from electoral politics can be a radical
one. For example, those people who are involved in new social
movement politics can see electoral politics associated with the
formation of a green party as a distraction from their serious and difficult
attempts at popular mobilisation for direct action. But such a decision to
abstain may not be radical at all. Those who see their work as being
primarily focused on pressure group activity may well view the
formation of a green political party as an unwelcome complication. It is
hard enough to try and persuade government to listen to environmental
concerns without this being construed as partisan activity on behalf of a
rival political party. There are also those who are members of existing

---

Box 5.1

Reasons for environmentalists to refrain from electoral politics

1 It does not challenge existing structures such as the bureaucratic organisation of the state or the profit system. Rather, entering election campaigns reaffirms the value of existing structures.

2 Focus on elections and dependence on sympathetic politicians does little to establish the social movement as a viable force outside the parliamentary arena. A basis for continuing struggle may not be established. Often after an exhausting election campaign, the movement virtually collapses.

3 The sense of personal responsibility for environmental problems is given away to elected elites.

4 Elected representatives, even those most responsive to community opinion, are still subjected to intense pressures to adopt anti-environmental or 'compromise' policies. Politicians are constantly influenced by industrial lobbyists and top state bureaucrats. A more pervasive influence is the requirement to maintain economic expansion in order to finance government programmes. Politicians cannot afford to jeopardise 'business confidence' by anti-capitalist policies. The key goal of political action becomes survival in office. For these reasons elites, including elected representatives, cannot be relied on to enforce environmentally sound policies.

5 Entering elections tends to polarise opinion on the environmental issue along party lines. Potential supporters in the party not endorsed become much harder to reach.

6 Election campaigning often depends on key personalities, either as candidates themselves or as charismatic campaigners. This dependence, plus the need to co-ordinate policies, maintain party unity and avoid doctrinal splits, tends to centralise power in the social movement itself, to reduce meaningful participation and thus weaken the base of the movement.

7 Strategies that do not depend on electioneering tend to be neglected.

(Adapted from Martin 1984: 111)

---

political parties and who have worked long and hard to have their party add green concerns to its traditional agenda. Such people can see no reason for creating new political parties that make their task harder and challenge those to which they already belong. So it is quite possible to reject the formation of green parties to be involved in electoral politics on conservative, reformist or radical grounds, without in any way compromising a commitment to effective environmental action.

Greening mainstream political parties

As noted in the previous section, some environmental activists are involved in efforts to green the existing political parties. These people make a realistic assessment of the chances that any new political party will succeed in displacing the traditional parties and see their efforts best directed at getting their party to take green issues seriously. This is no easy task since green concerns have to jostle with a whole array of alternative claims about what is important. Nonetheless, in some circumstances, in some places, existing political parties have adopted some green issues as part of their political calculations about how to position themselves to win in conflict with their political rivals. In essence, there are two ways in which political parties are 'greened': through the efforts of party activists and through political calculation and party competition. To understand the dynamics of this process, it is useful to consider a couple of examples.

The Democratic Party in the United States has had an uneasy relationship with the politics of environmental concern. A number of its key figures and presidential aspirants have embraced environmental issues as part of their campaigns without a great deal of success. The image of Michael Dukakis being pilloried by George Bush's campaign ads for his poor environmental record, although full of irony, was none the less effective in marginalising the issue. In that successful presidential campaign George Bush adopted some claims of environmental concerns, but there was little evidence that this influenced his conduct of policy making. However, Bill Clinton as presidential candidate chose Al Gore, who was well known for his attempts to make the environment an electoral issue (Gore 1992), as his vice-presidential running mate. Undoubtedly, Gore worked hard to get the Clinton administration to adopt better environmental policies, and evidence can be found to indicate a degree of success. As soon as he was elected, Bill Clinton reversed George Bush's position on the protocols and communiqués issued at the Rio Earth Summit. Nonetheless, Clinton's response to the increased anti-regulation militancy of the Republican majority in Congress saw a reversal on several important environmental issues, including restrictions on the destruction of wetlands by private property development. During the 1996 election campaign, Clinton manoeuvred between green initiatives and lifting regulatory restrictions: a political calculation determining the extent to which his politics could be described as having been 'greened'.

A very clear example of how political calculation and party competition can 'green' an established political party is to be found in Australia during the years of Labor government (1983–1996). There was an extended history of party competition and co-operation over environmental issues in Australia. For example, the attempt to sand-mine Fraser Island was opposed by the Labor government of Gough Whitlam (1972–1975), but the decision to ban mining was taken after a protracted public inquiry by the Liberal government of Malcolm Fraser (1975–1983). A public difference between the two major parties was a minor issue in the 1983 federal election, which saw a Labor Party success. At issue here was the willingness of Labor and the reluctance of the Liberals to use federal powers to override the Tasmanian State government on the issue of building a dam in the southwest rainforest 'wilderness' area of the state. So there was some important background for what happened in the mid to late 1980s.

The key to Labor's initial electoral success turned on its relationship with the trade unions and its ability to deliver sustained real wage cuts, undermining, to a degree, its electoral appeal. In these circumstances, Senator Graham Richardson took the environment portfolio and turned it into a significant force for environmental initiatives and a very effective part of a strategy to have the Labor Party re-elected (Richardson 1994). Richardson was the 'numbers man' for the New South Wales right, the dominant faction in the parliamentary party, a self-proclaimed realist, willing to do 'whatever it takes' to ensure Labor's success. What he did was to devise a 'green preference' strategy that would see those with strong environmental views vote in such a way that their preference votes would be transferred to Labor. His strategy involved making well-publicised, at times dramatic, important decisions to protect ecologically sensitive areas from economic development. This produced hostility in the business community and opposition from the Liberal Party but made it obvious that a Labor government would be better for the environment than its conservative opponents. In the 1987 and the 1990 elections, this strategy was a major part of the Labor campaign and its success. The strategy worked only up to a point, since the government did not adopt more than a 'green tinge' for its general policy making. Later, the priority of economic development was reasserted and environmental support either declined or was neutralised, such that in the 1996 election several environmental organisations were confused about the relative merits of the two parties.

These two examples show both the scope and limits of any strategy for greening existing and mainstream political parties. In certain circumstances, these parties will adopt more green images and environmental concerns, but the extent of this commitment will always be limited by political and electoral calculations. When the situation is propitious or demands it, these green moves will be dropped in favour of other policy initiatives and other views on how to win elections.

Green party experiments

Green political parties have been formed in many countries as a response to the experience of new social movement politics or the lobbying practices of environmental NGOs. The most successful green party was formed in West Germany in 1980 and achieved electoral success at the national level in 1983. The radical style of the West German Greens, with their commitment to participatory democracy, leadership rotation and gender equality, contrasted sharply with the normal politics of the rest of West Germany's conventional parties. Despite winning no seats in the 1990 're-unification' election, and a period of self-doubt over the strategies to be used, the German Greens are once again winning enough seats to be active players in both German and European electoral politics.

Although green parties have been formed all over the world, the experience of such parties in Australia, the USA, Britain and Germany is illustrative of the success or otherwise of these initiatives. In each country, environmental movements have used different strategies to deal with electoral politics and have had different degrees of success. It is this material that is the subject of evaluation for the rest of this chapter. But first it is useful to consider the contrasting experience of green parties and their basic characteristics.

Green partisan politics: case study summaries

For each of the four countries under examination we briefly discuss some specifics of their green partisan experience under five headings: (1) environmental consciousness; (2) political structures and arrangements; (3) political terrrain; (4) the state of green parties; and (5) green party highpoints.

Environmental consciousness

A high percentage of Australians have been measured as having environmental sensibilities. Most notions of environmentalism are post-materialist, oriented around wilderness concerns. Three times more Australians are members of environmental organisations than political parties.

A similar story can be told of the United States. In 1992, 70 to 80 percent of Americans called themselves 'environmentalists' (Easterbrook 1992: 26). North Americans, in general, are similarly dominated by the post-materialist values of the environment. The 'environment' has been dominated by non-human wilderness concerns and debates about pollution.

'Britain has the oldest, strongest, best-organised and most widely supported environmental lobby in the world' (McCormick 1991: 34, quoted in Richardson and Rootes 1995: 66). This claim, however debatable, has some grounds for support. British environmentalists, as described in Chapter 3, have a far greater range of environmental positions than those dominant in the Australian and US experience: from nature conservation to political ecology. Like many European environmentalists/ecologists, environmentalism in Britain is perceived by many as intensely *political*, and a sustained systemic critique is visible within and outside the movement.

Like the other three countries, West German citizens possess a high level of environmental consciousness. Many theorists argue along the post-materialist lines outlined in Chapter 3: that 'a new educated middle class who have grown up under conditions of relative peace and prosperity' are the base from which German Greens derive their support. Others argue that the unprecedented prosperity of West Germans led to unprecedented pollution, and this environmental degradation forced the West Germans' hand (Frankland 1995: 24–5). This is more closely aligned with the post-industrial arguments forwarded as explanation in Chapter 3. Whatever the origins of support, environmentalists in West Germany were intensely *political*. Not only was ecology a central plank, but so were participative democracy, social justice and non-violence. The environmental philosophy of Die Grünen, more than any other national experience, has shaped key tenets of global green electoral politics.

In the former East, environmentalists have emerged from other roots. The Alliance 90 group, formed for electoral purposes by East German Greens in the 1990 first all-Germany election, have a style of environmental consciousness vastly distinct from their West German counterparts. Jahn explains:

> Their protest was against the SED (Socialist Unity Party of Germany, the former East German ruling party) state, and they stressed the importance of 'dialogical politics'. This attitude implies, in sharp contrast to the West German Greens, an objection to being given the image of a party of the left. It was to be foreseen that in some states (for instance Saxony) Alliance 90 would even enter into a coalition with the Conservative Party (CDU) and cooperate with the authoritarian Ecological Democratic Party (ODP). In the negotiation process with the West German Greens, Alliance 90 demanded acceptance of the political system of the Federal Republic of Germany and the free market economy.
>
> (1994: 314)

Political structures and arrangements

In Australia, there are preferential voting systems for both state and federal parliament 'lower houses' (except in Tasmania). This has not led to green parties achieving representation in the House of Representatives. In the 'upper houses' (including the national Senate) there is proportional representation. All states have upper houses of 'review', except Queensland.

National elections are dominated by first-past-the-post electoral processes in the United States. Coupled with the immense difficulties of initiating a minor political party, this constitutes a major hurdle to green parties forming and becoming successful. Unlike most other parts of Europe, Britain also possesses a first-past-the-post system, designed for 'big government'. This, more than any other factor, limits green partisan success. Even in 'secondary' European Union elections it maintains this system.

Germany's electoral system, on the other hand, is dominated by proportional representation. Any party that achieves a 5 percent threshold enters the national-level *Bundestag*. Germany is also a federal system comprising *Landtages*. As aforesaid, federations also provide added advantages to minor parties. Consequently, these are 'more significant policy arenas than the local and regional councils of unitary systems' (Frankland 1995: 27). On the European Union stage, Germany's federated PR system is also readily apparent, giving it more access to far more direct electoral power than in any other of the three case studies.

Political terrain

Both the Australian Labor Party (ALP) and the Australian Democrats (AD) have sought to attract the green vote, in part waylaying the establishment of a national green party until 1992.

US Democrats, also, are clearly seen by many environmentalists as far more appealing than the Republicans. Other environmentalists argue that the Democrats do not go far enough to achieve adequate environmental reforms. The vast majority, however, do not involve themselves in third-party politics but pursue change through lobbying and, sometimes, direct action.

The Conservative Party held power in Britain from 1979 to 1997. There were attempts by the party to green its edges, but most eco-activists still see them as 'anti-environmental'. The Labour Party, also, has not addressed environmental issues sufficiently. The Liberal Democrats do make assertions that much of their platform is 'environmentally sound'. The European Union has offered Green Party activists another much-needed forum.

Currently, the German scenario is dominated by the issue of reunification. Its political success is integrally entwined with the ability of Greens from East and West to forge new, workable relationships that can be sustained. In addition, the alliances between the Greens and the SDP, which have formed several regional governments, must be seen to work.

State of green parties

Green parties have existed in Australia since the formation of the United Tasmania Group in the late 1970s. They exist at the local, state and national levels. The national Australian Green Party formed only in 1992; it formed so late for a variety of reasons. In addition to those listed above, there were intense rivalries between the national Greens and smaller green parties in both South and Western Australia. Also, there is a long-standing sentiment among many Australian environmentalists against participating in more mainstream political forums, like political parties.

Of the four case studies listed, the US history of green electoral politics is the poorest. Several green parties have been formed at local and state levels. At the national level, the 'Green Politics Network' was formed at a conference in June 1995. 'It seeks to group, under one spreading redwood, radical environmentalists and feminists, multi-cultural leftists, campaigners against the military budget, and that dwindling if hardy faction of citizens who still call themselves "socialist"' (Kagin 1995: 24). This network helped to foster the move to run a presidential ticket in 1996.

The British Green Party history is a relatively long and chequered one by green standards. The Ecology Party was formed in 1973 by a small group of activists from Coventry (Bramwell 1994: 121). According to Richardson, the British experience was dominated by the ideologically 'purer' *ecologists* (those who attributed intrinsic value to the Earth) rather than the *environmentalists* (those who simply cared for 'the betterment of human society') (Richardson 1995: 7).[3] In 1985, it was renamed the Green Party and now has numerous branches in both local and national electorates. The late 1980s to the mid 1990s were characterised by mixed results and internal politics (Hutchings 1994: 20–1). The internal disputes were not unlike the more famous West German disputes between the fundamentalists and the realists. Rootes observes:

> The dispute within the party is not about matters of environmental policy but about organisational structure. 'Realists' like Jonathon Porritt and Sarah Parkin assert that the party has squandered public sympathy by giving an impression of confusion and disorganisation and argue for a formally elected leadership, while 'fundamentalists' regard all talk of leadership as anathema and argue for maximum decentralisation within the party.
>
> (1995: 85)

After the success of the 1989 European election (14.9 percent), there was some hope that these electoral gains would somehow translate into similar wins in the 1992 general election. This did not eventuate, and the Green Party polled only 2 percent (approximately) of the vote. On one occasion, it suffered the humiliation of being beaten in a by-election by the Monster Raving Loony Party. Its political profile at both the local level and in the European Union is far more favourable.

Without reform of the electoral system, green parties will remain peripheral in British politics. This accounts for a number of central Green figures being active in Charter 88, the campaign for constitutional reform (Richardson and Rootes 1995: 86).

On 13 January 1980, a diverse alliance of activists launched the national West German Green Party: Die Grünen. 'They agreed that the pillars of the new party would be grass-roots democracy, social concern, ecology and non-violence' (Frankland 1995: 24). Early successes in the Landtage were built upon with representation in the Bundestag in the 1980s. This began in 1983, when they polled 5.6 percent of the national vote, which converted into 27 Green deputies. Four years later, this representation grew to 42 deputies.

In the early 1980s, Die Grünen referred to itself as a 'movement-party', understanding that there were major advantages to be had if some of its social movement characteristics could be held on to and further creatively developed in conjunction with a partisan wing. Partly as a result, even today the German Greens have problems finding day-to-day operational activists, as many of its supporters insist on devoting large amounts of their time to extra-parliamentary activities (Poguntke 1992: 250). On a more positive note, the movement-party aspiration challenged the way party politics was being played in West Germany. It held on to its grass-roots aspirations and based its intra-party procedures on *basisdemokratie*: 'Grass-roots democracy means the increased realization of decentralized, direct democracy. . . . We are determined to create a new type of party organization with decentralized structures that are designed according to the principles of grass-root democracy' (Die Grünen 1980, quoted in Poguntke 1992). Individual members and lower-level organisational units were given substantial power to participate directly in high-level decisions. The most famous example of these ideals was the 'rotation principle', which initially limited sitting MPs to two years in the Bundestag. This insistence on challenging the operational characteristics of the existing party frame-work led to many creative and sometimes destructive tensions within the party. Poguntke explains:

> In a nutshell, it can be argued that, organizationally, the Green Party represents the attempt to reconcile elite-challenging participatory aspirations with the constraints of party politics in a representative parliamentary system.
>
> (1992: 241)

This tension continued throughout the 1980s, often depicted as a conflict between two opposing factions: *Realo* versus *Fundi*. The Realos emphasised reform and playing party politics by the existing rules, including experimenting with alliances with the Social Democrats (SPD).

The Fundis involved themselves in extra-parliamentary activities and supported more radical declarations (Frankland 1995: 32). Towards the end of the 1980s, Die Grünen began to suffer, in part because of this feud.

Even more crucial than internal machinations was the political scenario provided by unification. The 1990 general election was evidence that Die Grünen and the East German Alliance had not successfully merged. Die Grünen polled poorly in this spirit of national unification (losing all its Bundestag seats), its election slogan demonstrating how out of tune it was with the country's mainstream sentiments: 'Everyone is talking about Germany. We are talking about the weather' (Joppke and Markovits 1994: 235). Its allies in East Germany won eight seats. After this disastrous election, in May 1993, two parties formed into one: Alliance 90/the Greens (referred to as The Greens). This alliance signalled the triumphant return of the Greens, polling 10.1 percent of the national vote in the June 1994 European elections, giving them twelve seats in the parliament. This truly was a major victory. Jahn contends:

> The future of Germany and Alliance 90/The Greens depends on resolving the symbolic struggle between competing answers to the problems facing a unified Germany.
>
> (1994: 317)

Interestingly, this struggle remained connected, for a time, to the battle of wills between the Fundis and the Realos. In the late 1980s, it had seemed as if the more radical Fundis held the upper hand. But since unification, the pragmatic politics of the Realos have seemed more in tune with the imagined free-market, sustainable development futures whose virtues were being extolled in the East. Many of the Fundis have now left the party battleground, and the Realos dominate.
Also, the much larger West German membership dominates in post-reunification politics. What is most fascinating is a new willingness to forge alliances with major parties (particularly the SPD), and this has actually led to the Greens sharing *government* in some Länder (states). The new era of actually governing will lead the Greens into many new kinds of policy dilemmas and other issues. For example, in 1996 Joshka Fisher, a Green member of the German parliament, voted to send troops to Bosnia as part of the Dayton Accord to protect the Bosnian Muslims: 'This has led to a split between pacifism and the prevention of genocide in the Green Party' (*New Perspectives Quarterly* 1996: 52–3).

Green party highpoints

There has been recent, moderate success on the Australian scene. Two Green candidates won election to the Senate in the 1996 federal election. Among these is the Green Party leader, Dr Bob Brown, made famous by his involvement in the Tasmanian Wilderness Society's successful bid to stop the building of the Franklin Dam in Tasmania's southwest forests in 1983 (Doyle and Kellow 1995: 133).

In 1989, also in Tasmania, five green independents were elected to parliament. They briefly formed a government by creating a Labor–Green Accord. During the mid 1990s, two West Australian green senators sometimes held the balance of power in the national upper house, and used this position to bargain successfully with parties both in government and in opposition.

In the the United States, highpoints are rare. In the 1996 presidential race, Ralph Nader, founder of the consumer movement in the USA, ran on a green party ticket in California. The importance of the California race may have pressured Democrat President Bill Clinton to advocate environmental causes more loudly.

There have been more numerous local victories, however, but few are better than those enjoyed by the New Mexico Green Party. In March 1994, Greens landed a seat on the Santa Fe City Council. In part response to this success, Roberto Mondrano ran for governor, and Lorenzo Garcia ran for state treasurer. Mondrano won 11 percent of the vote; while Garcia won a record-breaking 33 percent of the vote: 'more than any third-party candidate to run for state office in the last sixty years' (Skinner 1995: 14). The ousted Democrat governor accused the Greens of being 'spoilers'. The Greens' showing in the election gave them major-party

Plate 7 *Great Ravine, Franklin River. Courtesy of Chain Reaction*

status, which guarantees them a place on the ballot at the next general election, as well as giving them the right to hold primaries (Cooper 1994: 453).

In Britain, at the national level, the 1989 European Union election remains the pinnacle, with the Greens polling 14.9 percent of the vote. Unfortunately, this did not translate into seats in the European Parliament. There has also been considerable success in local government. There are anywhere between 100 and 300 councillors scattered across England (Hutchings 1994: 20–1), 80 percent of whom sit on the bottom tier: the local (parish, town or community) council. In some rural areas, green parties are more conservative than some of their urban compatriots.

Successes are many in the German context. For example, in the European Parliament elections the Greens won 8.2 percent of the vote in 1984; 8.4 percent in 1989; and 10.1 percent in 1994.

In the 1980s, consistently, Die Grünen polled well above the 5 percent threshold in national elections, guaranteeing multiple representation. After the disappointing all-Germany election of 1990, the unified Greens bounced back in the 1994 Bundestag elections, winning 49 seats (Jesinghausen 1994: 108–14).

By the end of the 1980s, in the Länder, Die Grünen held seats in eight of the eleven Landtage. These successes displaced 'the FDP as the third party, reducing the political power of the SPD' (Frankland 1995: 30), but the really important wins occurred in the 1990s. The Greens formed alliances with the Social Democrats actually to hold government in Bremen, Hesse and Saxony-Anhalt. Joschka Fischer, one of the members of the green 'inner circle', served as Hesse's environment minister (*The Economist* 1994: 51–2). In 1995, a coalition Green/SPD government was formed in Germany's largest state, North Rhine-Westphalia. This was seen, in part, as a testing ground for a similar alliance in the Bundestag (*The Economist* 1995: 45–6).

During the 1980s, these victories in the states aided, and were supplemented by, the election of some 7,000 Green local councillors.

Green parties considered

With the exception of the German Greens and the occasional state or local government party, environmental activists realise that it is very unlikely that they will be 'in' government, alone or in coalition. Why then are so many activists willing to give so much time and energy to the creation and running of green political parties? Box 5.2 lists some of the advantages that can flow from green party involvement in electoral politics.

Box 5.2

Advantages of green parties

1 To gain office as an elected representative.

2 To gain legitimacy, and to be seen as an 'authentic' political force by a wide audience *outside* of the movement.

3 To hold a balance of power.

4 To 'spoil'.

5 To make policy deals with more powerful parties.

6 To gain more access to the media. As the media is also dominated by electoral politics, environmentalists contributing to this sphere gain added legitimacy and their actions are more readily reported.

7 To count supporters, and more accurately judge the strength of opposition.

8 To gain financial rewards.

9 To lodge a 'protest' vote.

First, there is the possibility of gaining representation in an international, national, state or local representational forum. Such representation brings environmental concerns into parliaments, congresses and local councils. In this manner, environmental politics are either firmly placed on the mainstream political agenda or given wider publicity. This arrival of green issues on the electoral agenda signals the emergence of a new-found legitimacy. Environmentalists are now perceived as advocates of authentic concerns; no longer imagined by 'the public' as either peripheral or necessarily radical. This legitimacy may increase the popular appeal of environmental concerns and may attract more coverage from the mass media.

Enough green representation can give a minority party the balance of power in parliament and hence enhance its influence in bargaining over which party holds office and what kinds of policies are introduced. This situation has existed for over a decade in the Australian Senate, with greens, Democrats and independents holding 'the balance of power'. Sometimes this balance leads a major party to make a post-electoral alliance with a green party in a bid to form government. This occurred in 1989 in the Australian state of Tasmania, where the Australian Labor Party could only form government by forming an 'accord' with five green independents.

Often green representation has not been so significant. Instead, policy deals are done with major parties before electoral contests. A good example is the 1996 US presidential election, where Ralph Nader ran on a green ticket in California. Due to the electoral college system, California is the most important state for presidential candidates 'to carry'. Nader knew that he could not win the election, but he believed that he could spoil the Democrats' chances of winning if enough disillusioned 'left-leaning' voters deserted the Democrat camp. The last thing the greens wanted was to see a Republican president, giving the Republicans 'a full house': the House of Representatives, the Senate and the presidency. Instead, Nader wished to use his campaign as a lever to get environmental concessions from President Clinton. There can be no doubt that after Nader decided to run, Clinton became more active in environmental reforms, as with his decision to oppose the 'salvage logging rider', which had increased logging in public forests since 1995.

In both Britain and the United States, there is little prospect of greens gaining representation at the national level. Here, green parties use elections to attract media attention to their cause. In the 1989 European election, the British Greens, achieving a significant percentage of the vote, were separated out of the 'other parties' category in descriptive media analyses, and were highlighted as a fourth party next to the Conservatives, the Labour Party and the Liberal Democrats. This accomplishment saw a partial increase in media coverage of environmental issues, as the increased vote was interpreted by the media as a sign of extensive public support. The US Greens' 1996 presidential campaign was fought with a primary goal being 'to gain media coverage for their cause' (Kagin 1995: 24). There are also financial gains in playing party politics. Kramer writes about the West German context:

> They include reimbursement of election 'costs', public political 'education', political polling and research, political 'foundations', and so on. Of course only political 'parties' can join in this fun. The 5 per cent rule for parliamentary representation generally prevented smaller political formations from fully participating.
>
> (1994: 232)

The Green Party in British Columbia saw the possibilities of financial rewards. Party political contributions are tax-deductible in Canada, so it formed an Environmental Defenders' Fund with the slogan 'You pay $25 – Ottawa pays $75'. The money is used to support jailed direct-action forest environmentalists and their families by paying their fines and legal costs (Goldberg 1994: 13). Finally, minor parties can use elections as vehicles for protest. In Ireland, for example, many citizens vote for Comhaontas Glas in both local and European Union elections.[4] This support is not matched at national level. Many people feel that these secondary elections provide an outlet for a safe protest vote against status quo parties (Holmes and Kenny 1995: 224–5).

Determining why some green parties succeed and others fail

Why do green parties emerge and succeed in some countries and not in others? Most studies identify three types of factor to explain the rise and development of green parties. They are (1) the level of environmental concern; (2) institutional structures and arrangements; and (3) political opportunities.

Environmental consciousness

Obviously, there must be some level of environmental consciousness in a given country if there is to be an electoral response. In countries of Eastern and Southern Europe, for example, where environmental concerns have been interpreted (particularly since the 'transition' in the case of the East) as a luxury concern, there is less chance of developing a viable party response than in other parts of Europe. More often than not citizen support for environmental objectives only indicates a potential to be mobilised for electoral purposes. For example, Chris Rootes comments on European green parties in the following way:

> Two of the countries where environmental consciousness has been most consistently high and where the environment has ranked highly as a salient political issue – Denmark and the Netherlands – have produced only tiny and poorly supported Green parties, while in others where environmental awareness is less highly developed – such as Belgium and France – Green parties have been relatively successful. In Italy, where levels of both general concern about the national and global environment

> and 'personal complaint' about the state of the citizen's own environment
> were higher even than in Germany, the Greens have made only very
> modest electoral progress.
>
> (1995: 233)

The contention here is that although environmental consciousness is a
necessary prerequisite, the level of consciousness alone does not lead to a
successfully operational environmental/ecological party. Britain, the
United States, Germany and Australia all have high levels of
environmental support (as measured by numerous polls), but each
country has very different experiences with the development of green
parties.

Institutional structures and arrangements

The rules of the political game, as defined in the constitutions,
conventions or institutions of a particular polity, tend to determine what
can or cannot be done in a given context. Often, the design of a political
system is such that it is virtually impossible for new parties to enter the
mainstream political stage. For example, it is virtually impossible for
third parties to form and operate successfully in the USA. The laws in
most states make it difficult for third parties to secure a place on the
ballot by requiring large numbers of signatures, whereas the Democratic
and Republican parties are often granted automatic access. Further, the
public funding of campaigns is much more generous for the two major
parties. At a national level, for instance, third-party presidential
candidates receive money only after the election, if they have garnered
more than 5 percent of the vote, and only in proportion to their total vote.
Major party candidates, by contrast, get large election grants immediately
upon nomination (O'Connor and Sabato 1995: 461). Obviously, one way
around the American system is to be a very wealthy person, which is
how a presidential candidate like Ross Perot can emerge. Even with his
wealth he was unable to win, and this hints at the substantial problems
faced by any less well funded green political party.

In Britain, Germany and Australia, it is far easier to register as a political
party. In Britain's case only nominal deposits are required of
parliamentary candidates, with 'no restrictions upon individuals offering
themselves for election' (Rootes 1995: 238). Despite this ease of
registration, the British Green Party has achieved little success at the
national level. By far and away the most prohibitive institutional factor in

dampening both British and even American green party prospects is the nature of the electoral system in these countries. In addition, the relative success of the German and, to a lesser extent, the Australian green parties is, in part, the result of a favourable electoral system.

There are three basic electoral systems that need to be considered: first-past-the-post, preferential and proportional. In the first-past-the-post or winner-takes-all systems of Britain and the United States, it is simply necessary to get the largest number of votes cast in an electorate to win. There is no parliamentary representation for second place-getters and other minor parties.[5] First-past-the-post systems tend to produce governments based on a far larger percentage of seats held in parliament than the percentage of the popular votes cast. As a general rule, first-past-the-post works to entrench a two-party system and is believed to make for stable government.

The preferential system is used in the national and state lower house elections in Australia.[6] This system is not much more favourable for minor parties. A winning candidate must receive an absolute majority of votes (50 percent plus 1) in a single-member constituency. But few parties' candidates receive an absolute majority on the first count, particularly in a country like Australia, where every citizen must vote. So, the preferences of the other candidates are redistributed among the higher polling parties until one party crosses the 50 percent plus 1 threshold.

In this system, minority parties (including the Greens) do have a greater say, however indirectly, as to who governs by directing their preferences. Often policy deals are forged between prospective parties realistically vying for government, and minor parties pushing for, in our case, more environmentally responsible legislation. In the Australian preferential system, minor parties have not won direct representation, but their preferences have been significant and in some elections the two major parties have made 'bids' for these. Indeed, the 1990 federal election was won by Labor, at least in part, on the basis of the flow of environmental preferences.

Proportional representation is the system most open to Green political success and is used for elections in West Germany, New Zealand and to the Senate in Australia. To a large extent, this explains the existence of green representation in these countries' parliaments. This system gives minority parties a far greater chance of being elected (Doyle and Kellow 1995: 132). Proportional representation systems vary greatly but they share certain basic elements. Instead of the 'winner-takes-all' equation of the first-past-the-post electoral system, seats in parliament are distributed on the basis of the proportion of the votes won above a certain minimum level. Mostly, in proportional representation models, there are multi-member constituencies, and all that is needed to be elected is a 'quota' of the votes – and the threshold is far lower than 50 percent plus 1 of the preferential system. It is for this reason that defenders of the current British and American systems refer to their own systems as delivering 'strong government' and look with suspicion at proportional representation, arguing that it gives too much voice to minorities and favours coalition governments with the attendant prospect of instability.

The type of electoral system is the most pressing institutional factor dictating the existence and, to a large extent, the success of green parties in achieving representation in parliamentary systems. Rootes writes:

> In general, in countries with federal constitutions and proportional representation electoral systems, the institutional matrix is much more favourable for the development and success of Green parties, and for the development of mutually beneficial relationships between Green parties and the environmental movement, than it is in centralised unitary states with majoritarian electoral systems.
>
> (1995: 241)

Political opportunities, competition and context: the political terrain

Even a favourable electoral system and a high level of environmental consciousness does not guarantee electoral success. Being in the right place at the right time are critical determinants. Individual personalities also play a role, particularly in cultures that traditionally respect charismatic leaders. Sometimes, the chief factor that dictates success or failure may just be the standard of competency of party officials or the manner in which democratic needs are balanced by bureaucratic imperatives.

The previous sections discussed how some environmentalists worked with and within mainstream political parties to green their agendas and policies. In a country, state or locale where green concerns have been accommodated by established parties, there may be less scope for green parties to wave their alternative green banner; but if existing parties have failed to respond to environmental concerns, they may be more vulnerable to green electoral challenges.

Often electoral/political success may have little to do with the accommodation of green concerns, successful or otherwise. There may be, for example, a period of internal turmoil within one of the more powerful parties (sparked by ideological or strategic differences among its members), making that party less popular. Alternatively, it may be that economic or a multitude of other social scenarios have proved electorally unfavourable for a governing party. Whatever the reasons or non-reasons, mainstream parties are weaker during some elections, providing greater opportunities for green parties. On other occasions, the power of established parties may be overwhelming, shutting out any hope of minor party success.

The fate of the British Greens in the 1989 European election illustrates these points. Despite a poor showing in previous national elections, the Green Party gained 14.9 percent of the vote, which was the highest achieved by any Green Party in the European Union. This was achieved within the unfavourable first-past-the-post electoral system and as a result the Green Party won no seats. In 1989, a series of events had heightened public concern on green issues, such as 'a new awareness of global environmental problems such as the depletion of the atmospheric ozone layer and the "greenhouse effect" of carbon dioxide emissions' (Rootes 1995: 71). This was strengthened by more local concerns such as the quality of drinking water and the safety of basic foodstuffs. It should be noted that at this time the Conservative Party was emphatically dominant, that Labour had lost the national election badly and that the Liberal Democrats were not offering an effective alternative. A green vote was, at the very least, a very safe protest vote. Rudig and Franklin observe the following about this election in 1989:

> [the] special circumstances of the European elections appear to have favoured the Greens. Environmental issues had played an unusually important role in the preceding year, the center parties were in steep decline, and (more than any other EC country) European elections were not considered particularly important affairs; the political cost of voting Green was thus very low.
>
> (1992: 39)

The contingent character of political terrain is such that it has no final definitive shape. No doubt a different combination of contextual factors was at play in the 1994 European election, when the British Green Party averaged only 3.2 percent of the vote in 84 divisions (Rootes 1995: 253).

In Germany, Australia and the United States, different events have shaped the political terrain of separate elections. Initially, the West German Greens were dramatically weakened by Germany's reunification. At a time when nationalism and other issues took a front seat, Die Grunen polled poorly during the early 1990s. But since May 1993, a coalition between the East German Bundnis 90 and Die Grunen has led to renewed electoral success (Jahn 1994: 312). In the 1994 European elections, they achieved 10 percent of the vote, and twelve seats in the EU parliament.

Conclusion

There are three clearly different ways for environmentalists to play electoral politics. Choosing the first path, deliberately abstaining from party politics, promotes the possibility of longer-term, perhaps more profound, changes in the future. This is the political pathway of the outsider. It should be recalled that it is possible to reject forming green parties and taking part in elections on the grounds that a lobbying pressure group role is sufficient for the task. The second path leads to insider politics; to play with and within major political parties. Increased political access and the possibility of being part of a governing party may lead to favourable policy changes. The trade-off, of course, is that increased access is also achieved on behalf of such a political party into the affairs of the environmental movement. This may lead either to genuine reform or to the co-option of environmental goals. The final option is to form separate green political parties. This pathway is different from the others and it leads to two alternatives. It is possible to become so immersed in the party system that a green party begins to operate in much the same way as other established parties, with top-down leadership, centralised and hierarchical control of policy development, and a need to dominate the environmental movement so as to co-ordinate its actions with the party's electoral goals. The green party can also become committed to 'lowest denominator policies' designed to appeal to as many citizens as possible rather than being based on obvious and green principles. The alternative is to manage party politics in a manner which recognises that the green party is just one useful part of environmental movements, an electoral wing, not the evolutionary end-point for environmental dissent. This final pathway recognises that the different parts of environmental movements, political parties included, provide alternative venues for different types of environmental politics. Not all these are equally appropriate in changing specific political terrains. The greater the variety of ways of being politically active, the greater the chance of being politically effective in all the changing circumstances that confront environmental movements. Too great an emphasis on green party political activity may reduce the overall effectiveness of the environmental movement.

Further reading

Martin, B. (1984) 'Environmentalism and Electoralism', *The Ecologist*, 14, 3: 110–18.

Richardson, D. and Rootes, C. (1995) *The Green Challenge: the Development of Green Parties in Europe*, Routledge, London and New York.

⓺ Business politics and the environment

- ⦿ Business and the state
- ⦿ Rejection, accommodation and environmental business
- ⦿ Efficiency, profitability and technology
- ⦿ Managerialism and education

Introduction

Private business is one of the most important forces shaping the interactions between humans and the environment. This is not to say that private business is the sole force damaging the environment. Nor is it to assert that state-owned business has a more benign impact on the environment. Events in the Soviet Union and Eastern Europe should be sufficient to undermine such an interpretation (Carter and Turnock 1993). Nonetheless, private firms going about their normal business, unrestrained by too much government regulation, have delivered very significant amounts of environmental damage. Business has also been willing to invest considerable time and energy in resisting environmental regulations in order to protect its ability to continue doing whatever it wants without interference. As such, its political actions in defence of its ability to harm the environment need to be considered. But not all firms are indifferent to environmental damage. Parts of the business sector have worked tirelessly to develop and sell technologies that counter the environmentally harmful effects of industrial practices. (Companies developing pollution-control devices fall into this category.) Some firms choose to work with technologies that are inherently less harmful. Some have redesigned their production processes to reduce the adverse consequences. Others have 'gone green' in the marketing of their products. These variations in business' responses require some explanation.

This chapter starts by considering the power of business and the various political positions that different firms take towards environmental care and concern. It then considers the different ways in which business has responded to environmental criticism and the various arguments and strategies used by business to defend itself against both criticism and regulation. Included here is a discussion of business attitudes to sustainable development and the belief that increased economic and technological efficiency is a way of limiting the harm done to the environment. The chapter concludes with a set of case studies to illustrate the scope for business taking different positions on environmental questions, in different contexts, and with different management structures or philosophies.

Business power, environment and the state

What is business? Business is a way of organising economic activity (either in the production or circulation of goods and services) whereby private ownership of the means of production is combined with labour so that the resulting output can be sold to others to realise a profit. Characteristically, business is organised into firms or corporations that have a given legal form. This form both creates them as legal entities and provides protection against various kinds of legal action. For example, the limited liability company (which separates the owners and managers of a company from its corporate existence) was a very important invention that made it possible for entities to grow larger in size and more ambitious in action than was practicable for individual or family concerns. Companies and company structures may vary greatly in size and kind, depending on the economic and legal circumstances in which they operate. The basic goal of all private business is profit. If the money invested in a firm does not return a profit, its creditors and shareholders cannot be paid and the firm will cease to operate. In this case, the firm will either be declared bankrupt or will be taken over by another company. Private business lives in a world of potential or real competition. Individual firms confront rivals, and the presence of competition will force firms to seek to contain costs while maintaining profitability. It is certainly true that the size of major companies has increased greatly over the years and that, in some parts of the world (and some sectors of the economy), a particular firm may hold a monopoly (or be part of an oligopoly) that allows it to dominate a market. Monopolies and oligopolies certainly help to reduce the pressure of competition but they do not remove it completely. There is always the threat of a rival entering the sector, or of a technological change that undermines a dominant position. The scale of major companies has also changed as national economies have become increasingly global. Imperial expansion encouraged big firms to operate overseas, expanding their investment in international production and trade. Gradually these large firms have assumed a multinational or transnational character, linking stages of production and circulation on a global scale across national boundaries. For example, a sports shoe firm may have its headquarters and image firmly located in the USA, have production facilities in Korea and the Philippines, use materials produced in a whole range of other countries, and be selling the shoes into both suburban and outback Australia. The scale of operations encourages flexibility and grants firms the ability to switch between countries on the basis of any marginal advantage. All these factors have consequences for the way in which business deals with the environment and responds to both criticism and the prospects of environmental regulation.

Private business, given the pre-eminent role it plays in the economy, has numerous resources that it can use to protect its actions from criticism. Threats to reduce employment, investment or the ability to earn foreign exchange can all be used as power resources in battles over environmental concern. For example, if environmental critics seek to stop a mining project, the business concerned can always point to the project's economic benefits. These would include the number of people who will be employed in the construction and operation of the mine; the amount of money to be invested in the project; the commercial flow-on for other firms in the area (such as local suppliers); the value of the ore when processed or exported; and, finally, the amount of tax or royalties that will flow to the various levels of government. These are all significant economic contributions to a local area and a nation, so the firm's case can routinely be expected to be very persuasive. Indeed, so strong is the presumption in favour of private business and economic development that it will be relatively rare for business to deploy its power resources in obvious pro-development campaigns. In normal circumstances, the implicit power of business is sufficient to allow it to proceed with business as usual.

When considering the potential and actual power of business, it is important to consider the contrast with most environmental organisations. If the clash is between a large firm and a local environmental group (even if that group has links with national or international environmental organisations), then the business will almost always have more power resources to deploy than the environmental activists. More than that, the power resources to be deployed will be different, or at least skewed in distribution. In addition to the 'business' resources already noted, large firms have good quantities of money and professional resources that can be deployed in media and legal campaigns. By contrast, environmental organisations have to rely on their members for time, money and enthusiasm to mount and sustain any campaign. For environmentalists, co-operation and their ability to organise are important to offset these intrinsic sources of weakness. To be effective, organisation is needed to make a campaign work. For business, especially large businesses, organisation is an option rather than a necessity. Organisation here means a number of firms combining together to set up a specialised body to lobby government on their behalf and to co-ordinate a public response to, among other things, environmental challenges. Small firms are more vulnerable in any conflict (economic or political) and need greater organisation to be effective, even though competition makes it hard for them to organise in their own right.

The different role of organisation and the unequal distribution of power resources are not the only factors that govern what business has to do to protect itself from critics. The political regime is very important in determining what happens in environmental conflicts. Consider the bald difference between authoritarian/military rule and a more open democratic representative setting. The international oil company Shell has a relatively easy time defending itself against its Ogoni environmental opponents in Nigeria. Shell is so important to the Nigerian economy, and the revenue requirements of the ruling military regime, that there is very little doubt that the regime will stand by Shell and give it the protection it requires to continue operating. On the other hand, Shell operating in Western Europe is much more vulnerable, even when it is strongly supported by the relevant governments. Consider the political problem the company faced when it wanted to sink its Brent Spar oil rig in the North Sea. Greenpeace organised a well-publicised occupation of the rig and a strong media campaign presenting the case that it would be ecologically irresponsible to proceed with at-sea dumping. The British government, as mindful of the importance of Shell as the military regime in Nigeria, tried to support Shell's plans and give the proposed sinking good environmental credentials. Such government support continued as environmental protests mounted. In the end, Shell

Plate 8 *Graffiti by Bristol Greenham women in protest at BP's role in the nuclear fuel cycle through its giant uranium mine in South Australia. Courtesy of Helen Tann*

gave in and abandoned its cheap method of disposal. In the British/European case, the nature of the political process, its openness in particular, made it very difficult for even a powerful company like Shell to prevail. International protest put pressure on Shell, but it could not save the life of the Ogoni activist Ken Saro Wiwa. This suggests that some political regimes are more effective in their support for business than others.

Lest it be thought that the above contrast means that international business would always prefer to work with an authoritarian rather than a more democratic regime, it should be noted that authoritarian regimes can also cause problems for business. Shell in Nigeria is not free to do what it likes. It needs to seek the favours of the military regime for its continued presence in the country and mute any criticism it might want to make so as not to offend military figures. None of this either explains or justifies the poor environmental management of Shell's Nigerian operations or the environmental damage it continues to do.

It is important to recognise that authoritarian regimes and liberal-democratic representative regimes provide different settings and pose different challenges for business when confronting environmental critics. Authoritarian regimes tend to make life easier because such regimes are more likely to repress or restrict the actions of environmentalists, especially when they challenge development projects that are close to the hearts and pockets of the rulers. Liberal-democratic representative regimes provide much more scope for environmental activists and create much more authority for governments to intervene against specific proposals while maintaining support for private business as a whole.

With this background about the power resources of business and the importance of the regime setting, it is possible to turn to an examination of the internal divisions within the business community and the implications that these have for environmental performance and responses to environmental conflict.

Although business can be expected to unite behind the core requirement that all firms need to continue to operate (including a common desire to limit the implications of environmental criticism and concern), business can be quite divided over the environmental implications of any given project or the economic consequences of any proposed piece of environmental regulation. For example, a firm pumping waste into a river will have a different attitude towards moves to limit water pollution than another firm downstream that uses the water in its processes. Such differences are reinforced and reflected in different attitudes towards the environment and environmental concern. For heuristic purposes, the attitudes of business can be divided into three groups (see Box 6.1).

For the *rejectionists*, the case for environmental concern is, at best, a mischievous invention and more often a cloak to conceal the real intentions of those who are anti-business for a number of malign social purposes. Their goal is to protect business against this unjustified attack

---

Box 6.1

The three attitudes of business towards environmental concern

1 Some sections of business comprehensively reject the case for environmental concern (rejectionists).

2 Others are largely sceptical but have sought accommodation by making limited changes (accommodationists).

3 Some have embraced the themes raised in environmental critiques and have redesigned their processes to minimise damage to the environment (environmental business).

(Source: McEachern 1991: ch. 4)

---

on the rights of private property and managerial prerogatives. To the extent that environmental damage is recognised, it is justified by the economic development that results and the importance of economic growth (both for improving standards of living and, if need be, to address serious instances of environmental damage). The business community has always had its share of colourful characters who are frank in their rejection of environmental concern. This comment by Hugh Morgan, the CEO of Western Mining Company (one of Australia's major mining companies) is typical of the rejectionist position:

> The road to power for ambitious revolutionaries is no longer the socialist road. But the environmentalist road, today, offers great opportunities for the ambitious, power-seeking revolutionary. Environmentalism offers perhaps even better opportunities for undermining private property than socialism.
>
> (quoted in McEachern 1991: 110)

The *accommodationist* position has become the mainstream position for business as the levels of environmental concern have increased and consumers have registered some level of environmental commitment in the marketplace. In a sense, this position tries to maintain as much rejection as possible while accommodating key themes of environmental concern. Sometimes, the accommodation will be both strategic and rhetorical, as for example in the series of advertisements under the heading 'With Climate Change, What We Don't Know Can Hurt Us' published by Mobil to argue for a slow response to reducing greenhouse gas emissions (*The Australian* 6 Sept 96: 3). These ads do not reject environmental concern but seek to identify with it while arguing for a slow response, protecting the position of the energy companies and of energy-intensive industry. This position can also be seen in the various statements of the mainstream business organisations embracing sustainable development.

The *environmental business* position is not very common, and is frequently under attack from mainstream and rejectionist positions alike. Environmental businesses are thought to have conceded too much to the environmental critics and to undermine the positions taken by the rest of the business community. The approach of Anita Roddick and The Body Shop is one example of an environmental business. The Body Shop explicitly shaped its image and its product range around improved environmental practices and better relationships with its suppliers, while avoiding products tested on animals. Its attractiveness to its customers is based, in part, on an environmental and ethical criticism of its business rivals (who have responded with various attempts to discredit both the ethical and environmental credentials of the company and attempts to undermine its economic viability). Beyond the soap and cosmetic industries, it is now possible to find a range of companies that have redesigned their production processes and have then used this environmental redesign as a way of marketing their products.

The above three positions are not in themselves significant, as they merely describe the range of responses that business has taken to environmental concern. They become important when considering the political response that business makes to politicised environmental concern. Here these positions and the relationship between them combine to shape and constrain the actions that governments are willing to take on environmental questions. The balance between the three is not fixed and is conditioned by the political complexion of government and the options being presented on the policy menu. When governments hostile to environmental concern are in office, then the influence of the rejectionists is enhanced. When governments that are seeking to use environmental concern as an electoral resource are in office, green businesses have more scope for influence. Such shifts condition the response of the mainstream and change the substance that can be given to some very broad concepts seeking to link environmental concern with economic development. The most important of these concepts is that of sustainable development, which is discussed below along with other strategies used to defend the broader economic actions of business.

The greening of business?

To understand the business response to the rise of evidence of environmental damage and politicised environmental care, it is necessary to deepen our understanding of the three positions by considering three areas that can quite easily be interrelated by particular firms (depending on circumstances and political and economic opportunities). The first area concerns the way in which business in general defends its interactions with the environment against criticisms. The second considers the relationship of business to the gradual articulation of the concept of sustainable and ecologically sustainable development. The third looks at the way in which certain businesses at certain times have redesigned their operations to accommodate green criticisms and to find the economic efficiencies that are reinforced by such new forms of environmental calculation. These three areas can all be treated as part of an exploration of the extent (and the character) of the greening of business.

Business defends its relationship with the environment

When business is criticised for the harm it does to the environment, there are a number of propositions that it can use to defend itself. These are themes strongly stressed by the rejectionists and which form the bedrock of mainstream business efforts to limit the consequences of government regulation. The different themes are all arranged around a core claim about utility. Utilitarian arguments grew out of the forms of political economy dominant in Britain in the late eighteenth and early nineteenth centuries and are associated with the philosopher Jeremy Bentham. They hinge upon the notion that use and enhanced utility is the key to assessing value in an economic and social sense. In conflicts over the impact of business on the environment, it is the enhanced utility of production that justifies any harm done. The amount of damage is seen as far less significant than the enhanced utility of what business does. More goods are produced and circulated, standards of living rise, social wealth accumulates, and the ability of society to respond to environmental damage is enhanced. Here utility is linked to an account of the virtuous consequences of economic growth. It is important to note that this argument is usually constructed at the broadest level because it has a tendency to be unstable when confronted by the actual products produced and circulated. For example, the mining industry is willing to name products selectively and show their value. It will identify the role of coal in producing electricity and the good things that electricity does. It will also run through the range of useful products that are made from mined minerals, from battleships to aluminium foil. Nonetheless, justifying environmental damage is not so convincing when argued on the virtues of disposable polystyrene cups and packaging (recognised in practice by all who eat fast foods), disposable plastic shopping bags or the ever popular fluorescent shoe-laces. Trying to cite the utility of these products runs the risk of giving encouragement to a moral economy that is hostile to waste and the production of saleable but unnecessary products.

Linked to arguments about enhanced utility and the virtues of growth are claims about investment and employment. If environmental regulations or restrictions are imposed, then business can respond by claiming that both investment and employment will fall. Employment is used to justify whatever business is doing, since increased employment is a social good and unemployment is a social bad. Once again there is a dubious element in this claim, because much of what business does in its pursuit of increased profits also destroy jobs. For example, in forest disputes industry continually claims that logging restrictions will cost jobs. Environmentalists counter that the restructuring of the industry and the introduction of capital-intensive production technology has led to the closure of small saw mills and destroyed more jobs than environmental care. Here the rhetorical point of distinction is between the destruction of employment that limits profitability (environmental care) and the destruction of employment that enhances technical efficiency and boosts profitability. This invokes other arguments about economic efficiency and increased competitiveness, which are part of the everyday currency of contemporary political debates (see the discussion on technical and economic efficiency in the section starting on page 142).

When arguments based on utility and the virtues of the economy are exhausted, it is always possible for business to turn to the claim that business does not harm the environment any more than is strictly necessary. More than that, business cares for the environment as much as, if not (certainly in a more practical and effective sense), more than environmentalists do. Evidence for such environmental care may simply recycle claims about growth and the need to create the wealth necessary to fund responses to environmental problems. In other instances, the evidence for care can be based on the role of significant business figures in various conservation organisations and environmental campaigns. In Australia, for example, Sir Maurice Mowbray (a significant force in the mining industry) is noted for his role in promoting tree planting around mine sites as well as his involvement in the Australian Conservation Foundation. Similar examples can be found in the USA and Europe, especially now that various business organisations have set up specialist sub-groups to deal with environmental questions.

These three elements form the main ways in which business has responded to the challenge of environmental concern while seeking to preserve as much scope for continuing to do business as usual. At an organisational level, there has been a significant response to the concept of sustainable development because it suggests that investment and growth can be reconciled with environmental concern.

Sustainable development

The concept of sustainable development has had a relatively long evolution but became a central framework for discussing the interaction between business and the environment only in the 1980s (especially after the publication of the Brundtland Report, *Our Common Future*, in 1987).[1] The emphasis on sustainable development was reinforced by the publication of *Agenda 21* and the wide publicity given to the Rio Earth Summit in 1992 (see Chapter 8).[2] Initially, business was suspicious of the concept, which it interpreted as being hostile to its primary concerns and giving too much emphasis to evidence of environmental damage and potential threats to the environment. Nonetheless, business organisations rapidly came to understand that adopting sustainable development was an effective response to environmental criticism, not in the sense that it changed how business conducted itself so as to do less harm, but because it provided a rhetoric to protect the continuation of business as usual. For example, by the time the Australian edition of *Our Common Future* was published in 1990, the Business Council of Australia, the Confederation of Australian Industries, the National Farmers' Federation and the National Association of Forest Industries had all committed themselves to sustainable development (WCED 1990: 37–53). Similar examples can be found for the USA and Europe. In the USA, the President's Council for Sustainable Development (PCSD) included business figures who made a similar commitment to sustainable development, although it should be noted that this happened only when the final report was released in March 1996 (see PCSD 1996: 1–10; Cushman 1996: 1).

The business commitment to sustainable development has not meant great policy innovation, but when business needs to present a case for environmental concern then it can be packaged using this concept. For example, Shell was praised for its executive policy statement on sustainable development in the early 1990s (Callenbach *et al.* 1993: 36–7) despite the concurrent impact of its actions on the Ogoni in Nigeria. In the Australian case, the enthusiasm shown by business for sustainable development as a defence of its existing practices, while evidence of continuing environmental harm was accumulating, discredited the concept. It was in this context that the Labor government invented the ESD processes, adding 'Ecological' to the notion of sustainable development as a way of symbolising renewed concern and a new way of getting environmentalists involved in the process (see Chapter 2).

One of the clearest and most advanced statements of the business vision
of sustainable development comes in the publication *South African
Environments into the 21st Century* (Hunter, Siegfried and Sunter
1989; see also Cock and Koch 1991). The book was produced with
the involvement and support of the Anglo American Corporation (the
largest and most important company operating in South Africa). It was
published as a response to the growing evidence of the environmental
damage accelerated by the economics of the apartheid regime. Anglo
American had been committed to debating the economic options
facing South Africa and to exploring the requirements for a high-growth
scenario as a response to the challenge of extending the benefits of
economic development to the majority black population. Initially, it
was concerned that a high-growth scenario could only be achieved with
high environmental costs, and the book was an attempt to find ways
of showing that only high growth could deliver both developmental
and environmental benefits. The book consciously sets itself to
address an elite of both white and black opinion, to reject a focus on

Plate 9 *Community erected protest sign against nuclear waste dumping, Mitterteich,
W. Germany. Courtesy of Chain Reaction*

wildlife (the frequent focus of white environmentalism in South Africa) and to concentrate on the human future and the impact of environmental degradation. It is a confident rejection of apartheid with an optimistic account of development, provided high growth is associated with an environmental ethic and government provides 'the discipline in the market' (Hunter, Siegfried and Sunter 1989: 122). This is as far as business thinking goes on the question of sustainable development. It is an important alternative to the frequently expressed belief in unfettered free markets, which has dominated in the West since the early 1980s.

Efficiency/profitability/technology as environmental care

Business, as individual firms, is able to respond to increased environmental care in a way that is both good for its profits and for the environment while not 'turning green'. Some kinds of economic and technical improvements can lessen the impact on the environment of crucial production processes. For example, it is possible for buildings to be retro-fitted to improve their energy efficiency and lower their running costs. This can boost profits, lessen the demand for electricity and lessen the production of greenhouse gas emissions from power stations. It is a little thing but no less important for that. In a related way, firms can adopt zero-waste strategies and concentrate on redesigning production so that waste products (including heat) can be turned into inputs for other processes, improving efficiency and lowering the impact on the environment. The integrated production system introduced by Xerox in the USA is an example. It is not that these changes substantially 'green' business, but they certainly lessen the impact of business on the environment while maintaining a business-as-usual stance, which is attractive to managers and shareholders.

The fact that some businesses make such changes shows that it is possible for business to show care for the environment, up to a point. The fact that it is possible does not mean that all business can or will make the appropriate changes. Nor does it mean that the same firm operating in a number of different countries will not pursue high-efficiency solutions in one setting and low-efficiency solutions in others; protecting the environment in one and harming it in another. To understand the dynamics behind these business choices it is necessary to consider some contrasting examples.

Applications and examples

Mining in Papua New Guinea

Consider the case of gold mining on the Ok Tedi River in Papua New Guinea (PNG). PNG is a very poor country shaped by successive colonial regimes from Germany, Britain and Australia (Hyndman 1991). Its attempts to secure economic development have been conditioned by an urgent need to win foreign exchange, which comes most easily from mining and forestry. Mining ventures have proved to be quite risky given the natural and social conditions in which mining takes place. A large copper mine on Bougainville Island (run by CRA) produced major environmental damage and social disruption, which culminated in the closure of the mine when an independence movement launched a war against the mining company and the PNG authorities.

In other areas, high rainfall and remoteness combined to make mining difficult. It was in such circumstances that the PNG government sought to lock foreign investors into a mining project in the relatively remote area of the Star Mountains at Ok Tedi. High rainfall and unstable geological conditions combined to make the prospective copper and gold mine quite challenging. After trying a number of partners, Broken Hill Proprietary (BHP, Australia's largest mining company) came to own and manage the project through OTML (Ok Tedi Mine Ltd – a holding company in some kind of partnership with the PNG government). The economic importance of the mine is very great, providing a significant portion of the export earnings of PNG, a sizeable amount of revenue for the PNG government, and feeding economic and social development in the mine region along the upper Fly River. Along with this economic benefit has gone considerable environmental damage and social dislocation. The mine operates without a tailings dam, which would normally capture a lot of its waste. At one stage, there was an attempt to build such a dam, but it collapsed and has never been replaced. As a result, the tailings, crushed rock and eroded topsoil simply wash into the Ok Tedi River, increasing the sediment load. This leads to flooding and reduced fish supplies in parts of the river system. There has been a complex interplay between local land owners demanding compensation for the environmental damage and the PNG government's efforts to protect its project. BHP has expended considerable energy in defending its environmental record with claims about the naturally turbid condition of the river and the economic benefits of the mine (McEachern 1995).

From the perspective of the discussion in this chapter, what is important is the contrast between the operation of BHP in Australia with one set of environmental regulations and its operations in PNG, where weak government basically gives BHP a free hand to manage and monitor its own environmental impact. Mines have been run in Australia in ways that significantly damage the environment. For example, the Mount Lyell copper mine at Queenstown in Tasmania roasted its sulphur-rich ore with little regard for the consequent acid rain. As a result, the area around the mine resembles a barren 'moonscape', with little or no vegetation. This mine also operated without a tailings dam until 1994, so its waste polluted a significant area of Macquarie Harbour. Despite this record, it is not conceivable, even in the circumstances of a backlash against environmental care, that a new mine could be opened by BHP in Australia without a tailings dam to protect local rivers. Yet this happened in Papua New Guinea.

Business management education in North America

The long history of the Environmental Protection Agency in the USA, and the pressure exerted by numerous environmental campaigns, has made US business particularly sensitive to the challenge of politicised environmental concern. In this context, US business has developed a number of strategies to contain environmental criticisms through the use of political organisations and lobbying muscle. This section considers the reactions of US businesses that have responded to environmental concern by reviewing their production processes and seeking technical efficiencies that deliver both better environmental outcomes and improved economic viability.

The best illustrations of these gains come from a consideration of the recognition and treatment of waste. It is a commonplace to note that waste and pollution are basically resources that are not being used. So for example, the smelting of sulphur-rich ores releases vast quantities of sulphur dioxide and other sulphurous compounds. It has been common for these by-products to be released into the atmosphere as a waste, contributing to acid rain and damaging vegetation and soils. Under certain circumstances, smelting firms recognised that these sulphur compounds could be used as a raw material for other production processes (for example, they could be converted to sulphuric acid and sold as a raw material to fertiliser manufacturers). Capturing and selling these compounds would therefore increase the profit from each tonne of ore, as well as reducing the amount of environmental damage. In this sense, there is a more economic and ecologically efficient use of a non-renewable resource. A further example would be the capturing of heat released by one process so that it can be utilised by another process or firm. This synergy (one industry's waste product becoming the raw materials for another) can be generalised as a model for business responses to the environmental costs of its activities. The commercialisation and interconnection of by-product flows between firms can be proliferated seemingly endlessly. Such arrangements have been given the title 'industrial ecology' (Lowe 1996: 437–71).

An interesting question arises at this point. If such efficiencies and opportunities for increased profitability exist, why have they not been recognised until now? More importantly, what factors or forces can produce the changes necessary to make business see its waste either as a sign of inefficiency or as a potential source of revenue? One possibility is in the 'greening' of managerial education, so that decision makers will recognise such opportunities and act on them. The role of such education is very important in the USA, where there is a well-established tradition of MBA training for managers. This system is highly significant for three main reasons: first, many of the world's major international corporations base their head offices in the USA; second, many transnational firms recruit executives from the USA; and third, the American style of management training is being imitated in many other industrialised countries (Australia included). It appears that significant elements of environmental management have been introduced into the US managerial training system. For example, environmental case studies are being researched and taught at the Harvard Business School, the Haas Business School (University of California at Berkeley) and the Management Institute for Environment and Business (Washington DC). A number of books have also been published worldwide to extol the important business opportunities that lie in better environmental practices. These typically set out examples of good commercial responses to questions of policy design and waste management, promoting integrated production and environmental accounting (see for example Callenbach *et al.* 1996; Ditz *et al.* 1995; Fischer and Schot 1993; Hawken 1993; Lowe 1996). The International Standards Organisation is also developing specifications for monitoring and measuring the environmental performance of business. In Australia, the Commonwealth Department of the Environment has compiled an on-line *National Cleaner Production Database*, which cites examples of firms that reduced both costs and pollution.

It is not just managerial education that produces such a change in approach. Some modes of government regulation in conjunction with the pursuit of competitive advantage have also impelled closer attention to environmental standards and actions. For example, car producers in Japan, forced to adopt higher emission efficiency standards, have found that this helps them sell into the European and US markets as environmental concern and regulation tightens in those countries. Similarly, US chemical companies working on substitutes for CFCs were able to use negotiations over ozone regulation to give them a competitive advantage over international rivals. Even environmental impact assessment processes, entered into for public relations purposes, can play their part in making business sensitive to the advantages of changed ways of producing and using waste products, internalising forms of ecological rationality in the process (see Chapter 7).

Conclusion

Business power is one of the significant factors affecting both the degree of harm done to and the degree of care for the environment. With business pushing for a competitive edge and profitability, it can have strong incentives to reduce the amount of time and money spent on minimising the impact of economic development on the environment. However, if business can be persuaded of the need to factor environmental costs into its calculations, it can be a potent force for limiting or repairing ecological damage. Here, the pursuit of commercial advantage can sometimes drive business towards greener options and outcomes. The important question remains: under what conditions, in what circumstances and to what extent will business seek to harmonise its profit-seeking objective with protecting and improving the environmental condition of the world?

Further reading

Lowe, E. (1996) 'Industrial Ecology: A Context for Design and Decision,' in J. Fiskel (ed.) *Design for Environment*, McGraw-Hill, New York.

McEachern, D. (1991) *Business Mates: The Power and Politics of the Hawke Era*, Prentice Hall, Sydney.

Institutional politics and policy making: the greening of administration

- Institutions and bureaucracy
- Ecological and administrative rationality
- Governmentality
- Market-like instruments

Introduction

No matter how much environmental politics has focused on pressure group/NGO activity, the development of green parties and the use of environmental issues as part of the electoral contest between rival political parties or the activities of business, much of what happens in protecting or harming the environment comes from the management and making of policy in the hands of administration and bureaucracy. Whatever governments do, whatever laws or regulatory measures are introduced, these have their impact on the environment through the actions of various government departments and agencies. When considering the impact that politics has on the environment, much attention needs to be paid to the structure of administration, especially the interaction between departments and agencies with different responsibilities. Public administration is another site where the dispute between the demands of economic development and environmental concern is played out. The institutional design of the system of administration is important for what is done, since it can support or inhibit moves to take greater care for the environmental consequences of economic development. Some versions of institutional design may be more effective for having environmental concern written into the detail of a policy and into the practice of its administration. It is not just a question of the relationship between different government agencies but also a question of modes of rationality and the 'administrative mind'.

This chapter begins with a consideration of the nature of bureaucracy, the kinds of policy assumptions that are embedded in its normal modes of operation and the limitations these impose on environmental initiatives that come within its grasp. It then goes on to consider the 'battle' for the administrative mind and reviews the different instruments used to make and shape environmental policy, including an account of the currently fashionable use of market-like instruments.

The politics of bureaucracy

What is bureaucracy and what is its place in the political and policy-making order? This is an important question, and confusion often surrounds the answers. The formal descriptions of constitutions and popular political explanations can often be at variance with what happens in practice. Bureaucracy is the name given to those parts of the state that are principally charged with the administration of government decisions and programmes. This implies a simple and clear model of the state system and the interaction of its various parts. Government is at the centre of the system making the political decisions that define policy initiatives and the form in which they will be achieved, raising revenue and distributing resources through its various programmes and agencies. It is the responsibility of the bureaucracy to administer the decisions made by government. (The legal system, the police and the military also have their parts to play in the enforcement of government decisions and the rules of the system as a whole.) It is implicit in this description, and is occasionally stated as such, that the bureaucracy is not responsible for the ends of its activity, it merely neutrally administers the decisions of others as efficiently as possible within the resources made available by government. Such a view of a neutral administration can be misleading. It is certainly true that politicians make the key decisions, but the bureaucracy also has a part to play in providing information, advice and assessments of different policy options. As the 'Yes Minister' interpretation suggests, such advice can be quite significant in shaping, if not the decision itself, then the form of the decision and many of its finer details. Further, the bureaucracy might well have interests and concerns of its own that may not be the same as those of the politicians who are in charge. The tension between such divergent interests can have its impact on the making and administration of environmental policy.

This image of the decision/administration relationship is most appropriate to countries like the USA, Australia, Canada, Britain and other European countries. It is not clear how well this relationship model applies to countries with authoritarian and military regimes. Certainly, in countries where the pay of public servants is both low and irregular, it is unlikely that administration has the capacity to carry the decisions made by government. Other inefficiencies can appear due to different degrees of corruption, where enforcement becomes less systematic and impartial.

The foremost theorist of public administration and the politics of bureaucracy is Max Weber (1978). Weber is important for his foundational analysis of the 'rationality' of administration and its ability to produce an efficient linking between intention, means and ends. He is also important because later in his life, as he wrestled with the consequences of what he saw in administrative practice, he became an angry, depressed critic of what he described as 'the iron cage of rationality', the imperialistic application of administrative rationality to more and more aspects of life.

As far as Weber was concerned, administration achieved both its efficiency and danger from its ability to impose rationality on the performance of various tasks such that the personality, interests, associations and sympathies of the administrator would be stripped away and made irrelevant to the task at hand. The performance of a task could be codified in a set of rules and procedures, minimising discretion and maximising the certainty of performance.

'Ecological rationality'

The key assumption being made in Weber's analysis concerns a uni-directional conception of rationality. Although rationality has no content, its application renders predictable decisions that would otherwise be random or partial. 'Administrative rationality' of this kind (linking means to ends and dividing tasks into smaller, interlocking components so as to make decisions into rules and procedural boundaries) carries whole sets of assumptions about how to define the ends of administrative action, what counts as evidence, and how to measure technical efficiency in decision making.

If it is assumed that there is 'administrative rationality' of this kind, how does it relate to the 'rationality' of economic or marketplace decision making and how does it relate to the administration of environmental care? The relationship between 'administrative rationality' and the prerequisites of capital accumulation and economic growth is ambiguous. On the one hand, it is possible to find a positive relationship based on the conservatism associated with the practice of public administration. The social values implicit in the life-world of bureaucracy may embrace a given economic and political status quo and effectively serve the defined ends of that consensus. Hence, if that status quo is based on either a respect for private property (as in the USA) or respect for state ownership (as in the former USSR), in a context that favours economic growth over environmental concern, then bureaucratic conservatism will embody that ordering of priorities. In such circumstances, environmental concern, to the extent that it is given an administrative dimension, is likely to be handled in a way that favours this prevailing set of social/political assumptions. Radical and challenging environmental initiatives will be blunted and contained by the embrace of administrative rationality.

On the other hand, the logic of public administration does not rest on the same set of assumptions as that of entrepreneurial, competitive economic activity. Attitudes to risk and innovation can be quite different, and the whole spirit of pursuing ordered administration, detail and documentation could well act as a brake on the rationality of the marketplace. There could be antagonism between administrative and economic rationality. If that were the case, then there would be a basis for the effective administration of environmental concern. Indeed the same kind of administrative conservatism could provide an opening to environmental concern through extended/renewed regulation and supervision of economic conduct.

In exploring this puzzle, John Dryzek (1987; 1990) has posed a contrast between administrative and ecological rationality. Accepting Weber's account of administration as a way of conceiving rationality, Dryzek sets out a series of assumptions that take into account the principles of sustainability and what kind of ecological calculations these require.[1] On this basis, he then asks what it would mean for ecological rationality either to supplant or to supplement prevailing patterns of economic or administrative rationality.

At the heart of Dryzek's project are strands of optimism and pessimism. Pessimistically, Dryzek shows how strongly entrenched are both administrative rationality and its attendant forms of cost–benefit calculation. Optimistically, he shows how little needs to be changed before the organisational logic of bureaucracy starts to grind out better forms of policy making once ecological rationality gets into the process.

It is possible to make this argument because the way in which administration proceeds is compatible with the pursuit of newly defined social goals. Once the political process has placed environmental concern on the agenda and has enacted laws and regulations, the bureaucracy (which is neutral about ends) will render these administratively efficient. More than that, the way in which the bureaucracy supervises and regulates social conduct is compatible with the task of environmental regulation. Indeed the calculations, the range of information that needs to be gathered, assessed and applied, are but an extension of the normal mode of bureaucratic activity.

Establishing the balance between these two possibilities brings us to the debate over environmental intervention and the administrative mind.

The debate on 'the administrative mind'

In a major work edited by Paehlke and Torgerson (1990), the consequences of bureaucracy are probed for their impact on achieving environmental goals. Here the battle is between co-option and transformation. On one side is the view that once environmental politics gets into the administrative embrace any far-reaching, radical social and political implications are lost. More than that, the logic of administration means that environmental objections and environmental concern are broken down into administrative pieces and harmonised with the prevailing goals and assumptions of economic growth. Hence, there are accounts that show how environmental impact assessment (EIA) processes became devices for saying 'yes' to development with, at most, incremental adjustments to make environmental concern serve the ends of development. Hence, expressions of environmental concern are used to fine-tune and justify development plans, not to replace them with environmental care itself.

Against this interpretation, there is a more optimistic view that emphasises the corrosive effects of ecological rationality upon the ordered decision making of the administrative state. Instead of administration colonising the environmental project, the environmental project comes to colonise the administrative mind, displacing economic rationality by ecological rationality, which is then driven deeper into social and political processes by the routines of the administrative mind. Once again, EIA processes can be used to show how all this works.

Companies, to comply with environmental regulations, agree to undertake EIAs for purely pragmatic, instrumental reasons. Indeed, they may agree with the most cynical and shallow of intentions. Nonetheless, in the process of producing EIAs, and with the expectation that further EIAs will be required, the internal organisation of a company and of its calculations is altered. Part of the company may be given the task of producing or monitoring environmental performance. The public relations section may be given the task of overseeing the presentation of the company's case to a sceptical world. Some member of the board or senior management may be assigned responsibility for a newly created environmental portfolio. It is possible that within these slight, even cosmetic changes the seeds of an ecological rationality can be sown. For example, a company may, as a result of being forced to produce an environmental audit of its internal processes, come to recognise that it has been wasting valuable resources as pollution. As a result, a change in the production process could be ordered to increase economic and technical efficiency, which has the consequence of reducing a harmful impact on the environment (for a more detailed account of the economic benefits of EIA, see Thomas 1996).

It is always hard to trace such changes to a prime cause, for example the use of EIA procedures to justify development. It is slightly easier to show that such changes have been taking place in some companies in some countries at some times. It is not that ecological rationality has replaced administrative, economic and other forms of prevailing calculations, but even the addition of some elements of ecological rationality lessens the harmful impact that development can have on some aspects of the environment.

'Governmentality'

There is one body of recent theory that focuses precisely on the processes by which 'new' mentalities come into being and shape the interaction of government, companies and society. That is the literature on governmentality. 'Governmentality' was a concept proposed by Michel Foucault as a way of understanding the form of government that occurs in the contemporary era when coercion and regulation are not the only means used to produce ordered, predictable, manageable behaviour (Foucault 1991). This very basic idea has been taken up by others and used to understand the 'mentality' that makes people or an issue 'subject' to government. Miller and Rose (1993), for example, trace the way in which a proposed new accounting standard spreads and produces new modes of calculating economic performance, which then act to promote investment and growth. The introduction of these reforms does not rely on the direct regulation of economic activity by government but still produces regulation-like effects. Just as new accounting procedures can transform calculations, so can new ways of taking environmental considerations into account. An environmental problem, pollution for example, can be recognised. New modes of calculation and measurement can be introduced to produce the information needed for a solution to be found. A 'solution' then emerges from diverse places in the social world, not just from government. As a result, the 'problem' has been made amenable to decentralised forms of regulation.

One of the central themes of Foucault's initial lecture on governmentality concerns the place of the state in theorising both government and policy.[2] Foucault's purpose in seeking to theorise governmentality was to call into question the central role ascribed to the state in the understanding of government in the contemporary era. Foucault wanted to move away from a conception that saw sovereignty, government and policy making flowing down into society from the central command point of the state. Instead, Foucault wanted to suggest that the state was but one site for the promulgation and policing of policy and the conditions that governed the conduct of populations. His is an account of society treated as possessing numerous diffuse sites for initiative and resistance. In the field of environmental conduct, there is a sense in which we can trace through this diffuse pattern of policy making and policing the conduct of the population. Consider the situation in 1992 with the Rio Earth Summit, when the media was saturated with images and programmes of enhanced environmental concern, frequently focusing on the behaviour of individuals either as the cause of environmental problems or as the key to solving them. Business, advertisements, green groups, media outlets, politicians and other social forces all combined to send out messages about the need for individuals to change the way they behaved and to internalise and police themselves on the observance of these new rules for environmental conduct. This was not a question of a state edict or regulation but of diffuse sites of social power and influence propagating various messages that, as they intersected, mapped out new patterns for social conduct. This focus on the character of persons and the factors that govern their routinised and normal conduct complements and intersects with other social projects, changing the form of ecological calculations used in debating and assessing the environmental conduct of business.

Although there are problems with the governmentality literature (including an overemphasis on functionality, order and regularity as well as a failure to identify and analyse the mechanisms that produce new or unstable patterns of governmentality), the basic insight adds to our understanding of what has happened as environmental concern and regulation have become the subject of more extended debate. It certainly produces a different way of understanding the consequences of administration in the making and maintaining of policy.

Economic and market-like instruments

The possibility that environmental concern could re-enforce the urge for bureaucratic action has been noted by free-market economists, especially those associated with various right-wing, pro-business think-tanks. Their response has been to adapt their general critique of state intervention to this situation and to promote alternative, more market-like, approaches. To support their rejection of environmental regulation, there has also been a tendency to reject environmental concern itself as either fraudulent, misguided or based on far too complex and uncertain science to justify determined government action. It is instructive to revisit this critique of environmental regulation (crudely portrayed as a system of command and control) and the alternatives of 'flexible', 'market' mechanisms.

The use of direct government regulation to influence the impact of business (say) on the environment has been a common response to both particular and general environmental problems. For example, the decision to ban DDT because of its propensity to accumulate in the food chain is an example of 'direct' regulation. The government issues an order, a regulation, prohibiting the use of a particular substance except in prescribed, regulated or licensed circumstances. Such is the case with the use of ozone-depleting substances banned under an international protocol. For many years, efforts to reduce air pollution were based on state regulation of the amount of particular substances that could be released into the environment, with whole regimes of inspection and prosecution built around that regulation.

Despite the ease of promulgating regulations, there are problems with direct regulation, especially in the areas of policing compliance and in the unintended consequences that can follow from rigid enforcement. Often, regulation proves ineffective when enforcement is lax (which can be caused by a whole range of factors ranging from the regulatory agency being underfunded to political pressure to ignore regulatory breaches). Sometimes, it is claimed that the cost of regulation is not matched by the benefits that flow from it. Sometimes, the cost of compliance is high and firms may act to evade such regulations. Sometimes, the chances of being detected and punished for a breach are so low that there is little incentive to comply. All these factors undermine the efficiency of regulation.

The questions of rigidity and unexpected consequences are much more difficult and serious. An oft-cited example concerns a US decision to impose the fitting of 'scrubbing' devices to coal-fired power station chimneys in a move designed to limit air pollution in general and sulphur emissions in particular (Cairncross 1991: ch. 5). Although such a regulation falls equally on all coal-fired power stations, its impacts are costly and counter-productive because it handicaps high-polluting and low-polluting power stations alike. This forces up the price of power when, other things being equal, low-polluting power stations using low-sulphur coal could have responded more flexibly and more cheaply to the challenge. Such examples can be multiplied to show that bureaucratic regulation can be expensive and unable to deliver acceptable levels of environmental outcomes.

This account of bureaucratic inefficiency, tendentious though it may be, fuelled the thought of those wedded to neo-classical economics and inspired a whole range of measures to deliver the same kinds of environmental outcomes with greater flexibility and lower costs.[3]

The most common example used to illustrate this approach concerns strategies for pollution abatement. Instead of simply banning or limiting the emission of particular substances, a market for tradeable pollution permits can be created. First, some estimate is made of the total quantity of pollution being produced in an area or by a particular industry and a target is set for a lower level. Second, a number of permits are created that entitle each holder to emit pollutants up to a certain level (these entitlements add up to the total target level of pollution). Finally, these permits are then either issued, sold or auctioned off to polluters. Firms can then choose a flexible response to the problem of their polluting activities. They can simply buy enough permits to cover their current level of emissions because they find it cheaper than investing in new plant. They can buy fewer permits and reduce their emissions to the purchased level by investing in end-of-pipe pollution control devices. They can buy the permits they need now, invest in less polluting production technology later and resell surplus permits to other firms when their levels of pollution decline. The advantage of this scheme is that both the overall level of pollution is reduced and the price of the permits gives firms an economic incentive to reduce their production costs by reducing pollution. Some firms will find it cheaper to invest in pollution reductions, others will have to pay for more permits. In the end, all firms are required to pay for part of the costs of their environmental damage by buying permits. Cleaner firms pay less and dirty firms pay more (this at least partly internalises a previously neglected negative production externality). Another feature of this scheme is that it is possible to gradually reduce the aggregate pollution level further by either withdrawing permits, reducing each permit's pollution entitlement, or buying back permits to take them out of circulation. In this way, the scheme can further harness market forces and the price mechanism to the task of reducing pollution. This approach has the advantage of both providing flexibility and using the same kind of price signals that firms routinely use to monitor, plan and adjust their performance. It does not require a whole army of regulators to be effective. Firms regulate themselves and respond to the competitive behaviour of their rivals in the marketplace (a more detailed analysis of tradeable pollution permits is provided by the Bureau of Industry Economics 1992). These schemes have been tried in the USA with various levels of success. Rosenbaum suggests that the economic benefits of tradeable permits is nowhere near as high as some free-marketeers have hoped for (Rosenbaum 1991: 137).

This approach also identifies an absence of property as the source of environmental problems and proposes a market response. Following Hardin's argument in *The Tragedy of the Commons* (1968), it is argued that the lack of private property rights in a common-use situation produces indifference to the long-term economic and ecological health of an asset. Assigning private property rights to the asset generates both revenue to government and a material incentive for an individual or firm to see that the asset survives and is well managed. For example, when elephant numbers have fallen as a result of habitat destruction and poaching and the local population sees wildlife conservation as a threat to its existence, it has been proposed that local people be given property rights in elephants, which they can then use as an economic asset (DiLorenzo 1993). Locals can sell the right to hunt their elephants, sell their tusks for ivory (if the ivory trade is once again permitted) or keep the animals as a tourist attraction. In all cases, locals can make money from their property rights in the elephants. Further, locals now have an economic incentive to care for the elephants and to maintain their numbers. Whereas before, elephants were a dangerous nuisance, damaging crops, fencing and housing, they are now a valuable asset able to return an income to their owners. More than that, locals have an interest in protecting 'their' elephants against poachers. As a result of the granted property rights and associated incentives, it can be anticipated that elephant numbers will rise rather than fall, all other things being equal.

A whole range of these alternative instruments can be applied to a variety of environmental problems, including the application of 'green' taxes to change prices and send signals to encourage good environmental responses to ordinary everyday choices. Nonetheless, such instruments are not appropriate to all problems; some require state action for any chance of success. The best of the 'free-market environmentalists' recognise this and incorporate it into their response.

In the rhetoric of free-market economics, these initiatives are presented as alternatives to state regulation and as a way of avoiding the grip of bureaucratic supervision. At this point, it is important to note that this assertion is wrong. At most, these instruments are 'market-like' rather than market mechanisms. In their design, they harness the price mechanism to pursue designated environmental goals (such as reducing air pollution, improving water quality or preserving stocks of wild life) and they depend for their purpose and efficacy on political decisions and bureaucratic administration. These policy instruments are different ways to achieve the same kinds of goals as direct bureaucratic regulation. In such circumstances, the question of which instrument is preferred is not just a political question but is also a pragmatic one, a choice of means towards politically desired ends. In choosing between them, politicians and officials can make their choice on the grounds of technical efficiency and with due consideration of the kind of problem and the kind of solution sought. It is not a matter of state or market but of the form in which they are combined; it is not a question of regulation or deregulation but of changes in the overall mode of regulation. Even the most libertarian free-market solutions require the allocation and protection of property rights, as well as the policing of contracts and enforcement of agreements.

Consensus and participatory decision-making structures/strategies

One of the most enduring features of environmental politics and policy making is conflict. It does not matter whether the issue is pollution, changed land use, forest clearing, mineral exploration, dam building or quality of life, environmental incidents are frequently attended by groups of environmentalists in conflict with government and private business as they pursue development. Such conflict is completely ubiquitous and it is important. Government responses to environmental concerns are frequently impelled by the conflict that surrounds them. Often it seems as if, without conflict, governments would not respond to environmental problems. Given this, for governments and administration (those making and administering policy) conflict is as much a problem needing response as the environmental conditions to which it is attached.

How do governments respond to this dimension of environmental politics? The answer depends on circumstances, opportunities and the kind of government involved. If the government has a strong commitment to development and has either secure control, an authoritarian bent or a strong military base, it may simply ignore or repress dissent and allow development to proceed with whatever forms of environmental assessment are necessary to secure finance for the development project. The building of dams in Thailand and Malaysia, including the recent project to generate hydro-electricity on the basis of building a giant dam that displaces indigenous populations from forests, are examples.

If the government is involved in a liberal-democratic or other effective representational system, its options may be either broader or more narrowly constrained. It is still possible for such governments to choose to ignore environmental protests and opposition. For example, the Thatcher/Major Conservative Party governments had no trouble in using the police to remove protesters and to push through the construction of ring roads and by-passes. In Canada, the government in British Columbia used normal law enforcement to allow logging to proceed when protesters sought to blockade the forests. Governments in the USA and Australia have taken similar actions. Nonetheless, continual public activism and disruptions over environmental matters can be debilitating and politically costly. Under these circumstances, governments may seek to use a whole variety of devices to contain, incorporate or absorb conflict and hence minimise disruption.

The favourite device of governments of all kinds in many different political systems is the public inquiry. In Britain numerous Royal Commissions and public inquiries have been held on contentious issues that either already had engendered conflict or disruption or had the potential to do so. For example, Royal Commissions were held on the health risks associated with the Windscale Nuclear Power Plant (now called Sellafield), the location and type of reactors for Sizewell B, and for the location of various new airports. Indeed, many inquiries are held at all levels to deal with contentious and conflict-ridden environmental issues. In the United States, inquiries have been used to deal with chemical pollution, contaminated sites and wetland preservation. In Australia, the practice of holding inquiries was made routine through the creation of a Resource Assessment Commission (RAC), which was involved in making recommendations for managing conflicts over uranium mining at Coronation Hill (in the third stage of the expansion of the Kakadu National Park), export wood chipping from native forests and, later, on the management of the multiple uses of the coastal zone. In South Africa, the proposal to sand-mine dunes at Lake St Lucia, a Ramsar convention-listed wetland, was the subject of an extend EIA process that included vast amounts of public consultation and feedback (Preston-Whyte 1995; McEachern 1997).

All public inquiries, regardless of their title or varied modes of operation, have several features in common. They all empower some person(s) to hold an inquiry, to gather evidence (through submissions or by people appearing as witnesses), and to form an assessment on the basis of evidence and argument. This assessment is then forwarded to some form of governmental authority (a minister, a department, a local council), where the report is published (or suppressed) and the recommendations are either acted upon, amended, applauded, condemned or ignored.

As devices to diffuse or contain conflict, inquiries work in a variety of ways. Inquiries take the ragged and unpredictable forms of conflict and give them order and coherence. They require time, money and research effort to prepare submissions and, sometimes, to take part in the assessment of evidence and rival arguments. Frequently, environmental groups are short of these kinds of resources, and the effort of servicing the inquiry can detract from campaigns elsewhere, even when governments provide subsidies so that these groups can be involved. Further, there is a certain kind of authority that attaches to the 'independent', 'public' or 'expert' inquiry; this gives recommendations some standing that may sanction developments under conditions that would make continued opposition seem at best unreasonable and churlish. At other times, inquiries make findings that can be used to stop developments or impose conditions which make it likely that private developers will move their projects elsewhere. In Australia, the RAC inquiry into mining at Coronation Hill provided the context in which the government could make a decision that prevented mining. With proposals to sand-mine Fraser Island, an inquiry gave reasons for stopping the development. In South Africa, the massive inquiry into a proposal to sand-mine coastal dunes in the Lake St Lucia wetlands, coupled with the political process marking the transition from apartheid to majority rule, provided the context for a decision not to allow that development to go ahead. It is important to recognise that inquiries do not invariably favour approval or disallowal, but the decision to call an inquiry and the fate of its recommendations are always shaped by the political calculations of the government of the day.

Public inquiries and the like have a certain usefulness in dealing with some part of the continual incidence of environmental conflict. Nonetheless, conflict remains endemic to issues of this kind, and it increases the political costs for governments by introducing uncertainty into their political calculations, at least in those countries with open representational systems. Here it is not surprising to find that conflict-resolution mechanisms should be recommended for all kinds of issues as a way of keeping a problem from becoming politically significant. For example, in countries like the USA, Canada and Australia environmental conflict has accompanied the whole history of 'clear-fell' logging and the wood chipping of old-growth forests. In this context, it is not surprising to find attempts to discover cheaper and routine ways of trying to contain/diffuse conflict, largely on the basis of the sharing of knowledge. One of the most explicit cases of this can be found in the ESD process in Australia. As discussed earlier, the ecologically sustainable development process in Australia was a model of an inquiry that sought to incorporate the destabilising pressures from business, labour and some environmentalists.[4] In the working party on the forests (which did not include environmental representatives), a whole series of recommendations were made on ways of institutionalising and

Plate 10 *Alexandra Parade campaign – rally on Brunswick Street, Fitzroy 1995. Courtesy of Christiana Stergiou*

routinising forest conflict, taking it out of the forests and into information-sharing venues. Computer models, round-table forums, information sharing, forest tours and mediation processes were all recommended. It is assumed that if information is shared and all sides have a say then the chance of significant (politically disruptive) conflict is reduced and that old-growth forests can continue to be clear-fell logged in an ecologically satisfactory way.

Sometimes, such conflict-resolution procedures may help to contain conflict and, if they are effective, the views of environmentalists can be harnessed to the process of making a development project both more environmentally acceptable and more efficient. In these cases, the knowledge of environmentalists becomes harnessed to the process of enhanced, effective and politically acceptable development. Of course, such techniques cannot be effective all the time, in all places and for all kinds of environmental conflict, but the urge to find such devices will often be there.

Conclusion

The politics of bureaucracy, the institutional design of the state and the variety of policy instruments chosen to respond to both problems in the environment and increasing levels of environmental concern vary from country to country and from time to time. Changes in design and preferred policy instruments do not seem to be a product of increased skill in the making of environmental policy or a result of increased awareness of the interconnection between economic development and environmental consequences. Rather, the structure of institutions, the procedures for their co-ordination and the choice of policy instruments seem to be more influenced by a broader debate about the character of government and its impact on economic growth. Hence, when the mood changed against state ownership, regulation and intervention, there was a corresponding change towards market-like policy instruments. No matter how strong the preference is for markets, politics, government and bureaucracy set the frame within which environmental policies are made and evaluated.

Further reading

Dryzek, J. (1987) *Rational Ecology: Environment and Political Economy*, Basil Blackwell, Oxford.

Paehlke, R. and Torgerson, D. (eds) (1990) *Managing Leviathan: Environmental Politics and the Administrative State*, Belhaven Press, London.

Weber, M. (1978) *Economy and Society: An Outline of Interpretive Sociology* (edited by G. Roth and C. Wittich), University of California Press, Berkeley.

8 The global dimension to environmental politics

- Global issues and forums
- UNCED and Agenda 21
- Population control and carrying capacity
- Climate change and ozone diplomacy

Introduction

In producing a comparative account of environmental politics, there is a tendency to replicate the assumptions and limits of the nation-state. Examples are drawn from events within particular countries and the similarities and differences between them can be noted and assessed. In considering the nature of environmental politics, it is important to notice the extent to which the division of the world into competing or co-operating nation-states is a problem with implications for understanding and responding to environmental problems. Often, an environmental problem has either a global/international dimension or its effects and solutions cross national boundaries.

A few simple examples can illustrate this point. If forests are cleared in Nepal, this affects the volume of water that runs off the hillsides into the rivers. An additional burden of silt from the consequent soil erosion results in greater flooding in Bangladesh. This is an environmental problem that requires co-operation across a national boundary for its resolution. Without successful international diplomacy, the environmental problem cannot be solved. This cross-border dimension may be present in a whole range of situations. Sulphur emissions from coal-fired power stations in Britain may be linked to acid rain in Europe and increased environmental stress on native forests. Other environmental problems are global in a more significant sense. The release of ozone-depleting substances into the atmosphere from a whole range of places across the globe can cause a change in the upper atmosphere that is then experienced as an environmental problem in a number of countries. None of these can solve the problem by any actions that they take on their own, because their individual contributions will be insignificant if others continue the unrestricted use of these chemicals. The enhanced greenhouse effect generates a similar global environmental problem.

There are three ways in which environmental problems raise the need for international responses. First, there are problems shared by a large number of countries across the globe. Questions of pollution, soil degradation and forest depletion are of this kind. Since these problems are shared, there is some point in holding international conferences and agreeing on broad policy frameworks. Much of what happened at Stockholm in 1972, in the production of the World Conservation Strategy in 1980 and the Brundtland Report of 1987, focused on sharing an appreciation of common problems and the array of policy instruments that can be used to respond – a sharing of awareness and experience. Second, there are problems that may be concentrated in some countries but which carry global implications. Energy and resource use in the United States (as the largest consumer) carries such implications, since it acts as a constraint on the standard of living in other countries and limits their ability to have access to equivalent quantities of resources at similar prices. Population issues also fall into this category. Again, these issues can be addressed in global forums but they cannot plausibly be solved there. Third, there are those issues that are either cross-border in character or are genuinely global. In the case of cross-border issues, diplomacy is required. Genuinely global problems require genuinely global solutions.

This chapter begins with a consideration of two global forums: that at Rio de Janeiro regarding environment and development (1992) and at Cairo regarding population (1994). These forums, particularly the Rio Earth Summit, tried to deal with environmental problems of all types, cross-border, international, global and shared experience. An exploration of such global forums should indicate both the strengths and weakness of the summit conference approach. Having examined global forums, the chapter then explores two examples of global diplomacy in response to global environmental problems: ozone depletion and the enhanced greenhouse effect. Once again it is possible to assess the circumstances under which global diplomacy can be effective.

The Rio Earth Summit and Agenda 21

Numerous global conferences have been organised around environmental themes since the first in 1972 in Stockholm. Many of these have been associated with the United Nations Environment Programme (UNEP) as it has provided personnel, funding, information and education to support these global initiatives. The Stockholm conference was, in a very real sense, a preliminary affair (Ward and Dubos 1972). It was more involved in promoting awareness and outlining a timetable of activities (which culminated in a big environmental conference in 1992) than in drafting and issuing some broad policy framework in the shape of the usual communiqué. UNEP was subsequently involved in the process that produced the World Conservation Strategy (1980) and the Brundtland Report (1987), as well as promoting the research and diplomatic initiatives that eventually produced the Montreal Protocol on ozone-depleting substances. It has been a major player in seeking and promoting both global awareness of environmental issues and global responses to environmental problems. The 1992 conference was to be its major achievement.

The Earth Summit, or more formally the United Nations Conference on Environment and Development (UNCED), was held in Rio de Janeiro in Brazil from 3 to 14 June 1992. Jordan writes:

> It is difficult not to review Rio without repeating some of the 'key' (but now hackneyed) facts: 'the Summit was attended by 130 heads of state, 1500 non-governmental groups and 7000 accredited journalists'; 'it was the largest high-level intergovernmental conference ever held'. But, 'Rio' was only the most visible tip of a much bigger iceberg of activity. Prior to the Summit, there had been almost two and a half years of international organisation(s) at a variety of fora dotted across the globe: preparatory committees; intergovernmental negotiating committees; meetings of the G77 (developing) states. Eventually, these negotiations culminated in the agreement of two opaquely worded 'hard law' conventions on climate change and biodiversity protection, and three pieces of international 'soft law'.
>
> (1994: 160)

Some degree of institutional redesign was also accomplished at Rio with a 'reconfiguration' of the World Bank's Global Environmental Facility (GEF) and the establishment of the UN Commission on Sustainable Development (UNCSD). The largest document released at Rio was *Agenda 21*, an account of what needed to be done to secure a global future into the twenty-first century. Rather like the documents that went before it (the *World Conservation Strategy*, *Our Common Future* and the initial Stockholm conference proceedings), *Agenda 21* combines an account of environmental and ecological damage with a suggested set of priorities and policy responses to achieve some form of sustainable development. Producing *Agenda 21* was an enormous and frustrating exercise in drafting, negotiating and redrafting to find an acceptable version that could be authorised by the Earth Summit. Indicating the character of this compromise and consensus is the first principle, which begins: 'Human beings are at the centre of concerns for sustainable development' (UNCED 1992).

The politics of the Rio Earth Summit are worth evaluating. There was a sense in which the conference and *Agenda 21* reinforced understandings of the global dimension of many environmental issues, and this was reflected in the popularity of terms like 'interconnectedness', which were freely added to the public vocabulary. It was broadly accepted that the environmental problems of a country or region can have an impact, either directly or indirectly, on the rest of the inhabitants of the globe.

This recognition of interconnectedness and the globally shared character of environmental problems influenced the policy initiatives of the governments of the more affluent countries. In the past, some environmentalists like Ehrlich, some deep ecologists and the neo-Malthusians defined population pressure as the key global environmental issue and a threat to their parts of the world. They sought to defend their environments by separating out their countries from any global context and sought to confine the consequences of population growth to the poverty of people in remote parts of the globe. To the extent that they had a global perspective, it was to pressure poorer countries to take efficient methods to lower the birth rate. A recognition that the biosphere had to be shared, that the demands of economic growth in one part could endanger environmental conditions in another, forced a rethinking of this perspective. Some environmental activists and some governments sought to use these global forums as a way of pressing for rules of conduct that would protect their countries from the harmful consequences of bad practice in other parts of the world. The prescriptions for sustainable development can be interpreted in this way.

This is not the only 'global' response to this challenge. Business, fully aware of the global scale of its operations and of the threat posed by both environmental damage and environmental concern, spent a good deal of effort to see that the policy framework promulgated at Rio was compatible with its broad interests and requirements. The terms in which sustainable development was defined in *Agenda 21* echoed the way in which business organisations think about acceptable policy settings. Nonetheless, Rio was not the only forum within which business operated to achieve a global framework for its actions, a framework that could limit the kinds of environmental action that could be taken against it. Certainly, Rio was important but not as important as the protracted negotiations that produced the World Trade Organisation (WTO) and the new General Agreement on Tariffs and Trade (GATT). GATT has clauses that limit the circumstances in which environmental regulation can be imposed if, by acting, there is an impact on trade. It is possible for a country to regulate business interactions with the environment, but it is not possible for it to do so in such a way that it has an impact on trade. The government of one country can challenge the validity of environmental regulations and restrictions in another country and have them rejected. The acceptance of the 'environment' as an issue by the World Bank and the International Monetary Fund (IMF) is to be understood in these terms, as the product of the changes in the institutional setting, partly as a result of the United Nations Conference on Environment and Development and partly following the creation of the World Trade Organisation.

Although *Agenda 21* sets out guidelines for achieving a form of global environmental protection, the negotiations that produced it marginalised a number of NGOs and excluded a whole range of issues.

Much was made at the conference of the inability of government and, particularly, non-governmental representatives to address substantive issues that they had deemed important. Wagaki Mwangi, a representative of the Nairobi-based International Youth Environment and Development, expressed these complaints in this way:

> Those of us who have watched the process have said that UNCED has failed. As youth we beg to differ. Multinational corporations, the United States, Japan, the World Bank, the International Monetary Fund have got away with what they always wanted, carving out a better and more comfortable future for themselves. . . . UNCED has ensured increased domination by those who already have power. Worse still it has robbed the poor of the little power that they had. It has made them victims of a market economy that has thus far threatened our planet. Amidst elaborate cocktails, travelling and partying, few negotiators realised how critical their decisions are to our generation. By failing to address such

fundamental issues as militarism, regulation of transnational corporations, democratisation of the international aid agencies and inequitable terms of trade, my generation has been damned.

(quoted in Chatterjee and Finger 1994: 167).

Why was it that the vast array of delegates, official and unofficial, who gathered at Rio had so little impact on what happened and the content of the documents issued? Earth Summits, even where they involve a mass of delegates, are just like all other summits held between world leaders. What is achieved, signed or published is usually the product of extensive diplomatic negotiations, which take place behind closed doors and before the conference is held. At most, summit talks clarify and resolve a limited set of outstanding, difficult or sensitive issues. The success or failure of UNCED had very little to do with what actually transpired at Rio; the broad details of what was to be agreed had already been determined by the more powerful states attending, the conference secretariat and powerful business organisations. The conference itself was largely the 'show' used to 'launch' the product, *Agenda 21* and the various conventions and agreements.

The United States largely shunned the Rio process and the documents that were born there. Under the leadership of President George Bush, the United States was largely interested in promoting increased protection of tropical rainforests, while at the same time seeking to weaken agreements on climate change and biodiversity in its own favour. In fact, the USA was the only industrialised country not to sign the biodiversity convention. The US stance was not determined by the politics of the Earth Summit and diplomatic negotiations between countries, but by the internal dynamics of US domestic politics and White House political calculations. Schrader writes:

> To many participants, Bush seemed paralyzed by the dominant impulses of his Republican party. The internal battle over the treaty had been fought and decided months before the conference. David McIntosh, the executive director of the President's Council on Competitiveness . . . had ruined the Rio treaty and the EPA's chances for success by savaging the policy in an April memo. He warned that the treaty would impair the ability of American corporations to protect their biological inventions and technologies overseas.
>
> (1992: 431)

Bush's concern with his image of 'toughness' and 'leadership', the upcoming presidential election, and the fractious nature of the Republican Party shaped the way in which the USA both supported and limited initiatives taken at Rio. Once Bill Clinton had been sworn in as president, domestic considerations were again behind his decision to sign the Biodiversity Convention. In practice, the change was no more than symbolic. There was no change in the way in which the USA defined its interests on these global environmental issues, just a change in the ways in which those interests would be presented and defended.

All international summits and conferences are full of hype. Expectations of dramatic happenings and outcomes are created, only to be disappointed by the mundane proceedings. For a time, the world's media came to be focused on the environment. Environmental activists and government players gathered together, networks of contacts were created and the conference closed. The circus moved on. Rio was an overblown affair (both in terms of costs and expectations), and its achievements were modest and could have been attained more cheaply by conventional means. As a result, conferences of this kind became relatively discredited. The set agenda of mega-conferences would continue (the Population Conference in Cairo, the Women's Conference in Beijing and the Habitat Conference in Istanbul), but it was unlikely that an Earth Summit like that in Rio would be held again. Sandbrook writes:

> the world has learnt that there is no easy set of solutions to be had at an international level to many of the world's environment and development ills. National and subnational administrations hold the key. If they respond to this challenge, there will be progress. . . . The optimists would argue that there is now on the table a new and powerful agenda relating to the long-term welfare and a protected planet. The question is who will pick it up and do something about it.
>
> (1993: 30)

The Cairo population conference

> The facts speak for themselves. We have one Earth. One life support system. And it is shrinking. Over the past year alone, the global population has grown by 93 million people, bringing the total global population to about 5.7 billion. . . . This is an important meeting at which the stakes are high. They are nothing less than a sustainable future for our children and our children's children.
>
> (Dowdeswell 1994)

Another UN 'mega-conference' took place in Cairo in 1994. One hundred and seventy nine countries and thousands of NGOs participated in creating a broad-ranging 'Programme of Action' that placed the population issue at the centre of social development (Johnson 1995). To understand what happened at Cairo, it is necessary to reconsider the 'population debate'.

Chapter 2 compared deep ecology, social ecology, socialist ecology, eco-feminism and some other more radical eco-political philosophies. The issue of population was introduced there as a key 'litmus test' dividing deep ecologists and those radical ecologists of a more anarchist or leftist bent. Deep ecologists are just one group of environmentalists, some of whom could be classified as neo-Malthusians. Thomas Malthus, the 'dismal parson' (see Chapter 2), presented his 'population principle' 200 years ago. He argued that the Earth's food production resources would increase arithmetically, while the Earth's population would grow geometrically, so population growth would outstrip the ability to provide food. Mass poverty and misery would be the result. This line of reasoning has been used to justify birth control programmes and to urge no response to poor people in need as a result of famine and drought. It was an extremely popular argument in the 1960s and 1970s, championed by both Garrett Hardin and Paul Ehrlich. Their argument begins with the assumption that there are already too many people consuming the Earth's limited resources. As a result of decreased infant mortality and increased life expectancy spreading across the globe as a consequence of improved health care, the world's population will increase rapidly. A 'population bomb' is set to explode. In the literature, this bomb was primed by developments largely in Africa, India and China, and the question of resource use was largely ignored. Human numbers were the most dangerous indicators of environmental degradation on the planet. With such a narrative other questions emerge: what is the Earth's 'carrying capacity'; and who should have access to increasingly scarce resources? David Pepper writes:

> In this view 'ecofascism' is the adoption of the lifeboat ethic whose basic tenet is that because of absolute constraints on the amount of resources which are available there cannot be equality of access to those resources unless the numbers wanting access are greatly reduced and the responsibility for eliminating 'overpopulation' lies with those who are 'causing' it by breeding too fast. Any attempts by those possessing resources to redistribute them to those who are still 'irresponsible' enough to 'overbreed' are themselves responsible, for they merely result in more overbreeding.
>
> (1984: 205)

Plate 11 *Reclaimed gully, Son valley, north-central India. Courtesy of M. A. J. Williams, private collection*

Critics of the lifeboat ethic, like Pepper, argue that it has strong conservative, right-wing and, sometimes, fascist political origins. Pepper claims that these views have been taken up by the deep ecologists, who are now among the most vociferous advocates of population control. He notes that it is rare for those who advocate population control to address the overconsumption of resources by the wealthier nations and other aspects of the maldistribution equation. Pepper's eco-socialist position simply does not rate population control high on the eco-political agenda. To both social ecologists and socialist ecologists the population question provides a 'smokescreen' to avoid far more important issues of adequate welfare, shelter, education, emergency relief and equity.

The population question has not only been the subject of debate between rival schools of environmental activists. The obvious connection between the size of a population, economic activity and the standard of living is a commonplace part of studies of the history of national economic development. Further, the rate of population increase can be shown to fall as levels of security and economic welfare rise (hence the present demographic studies, which produce different models to predict the point at which world population will stabilise in the twenty-first century). UN and other international development agencies, as well as the governments of poorer countries, spend considerable time and effort on their strategies to reduce birth rates through the use of contraceptives as part of family planning strategies aimed at improving both the standard of living and the quality of life. The Cairo Population Conference came out of a convergence of arguments about both economic development and the environmental consequences of population growth.

In the early days of the Cairo conference, a coalition of Vatican and conservative Muslim leaders condemned the United Nations and its population programme as 'a Western imperialist assault on the family' (Heschel 1995: 15). The Vatican actively defended its anti-abortion and anti-contraception position. Interestingly, contraception was not officially opposed at the conference. The media, however, portrayed feminists attending the conference as the 'winners' in managing to reframe the debate to suit their purposes. Heschel explains this line of reasoning:

> Overpopulation, the conference declared, is an aspect of a larger problem: the maltreatment of women worldwide. From the outset, the modern feminist movement has argued that a woman cannot control her life if she cannot control her fertility. At Cairo, this was reformulated: unless a woman has a life of her own, her fertility will be out of control.
>
> (*ibid.*: 18)

Indeed, defenders of the conference believe that the emergence of this perspective was a turning point. Women were now perceived as the 'agents of change' in the goal of population stabilisation and economic development. 'Empowering women' as a focus of international aid policies and of national government initiatives would certainly change both the population equation and the approach to economic development. The difficulties of the task should not be underestimated, or the degree of domestic and political opposition. Some critics of this conference have pointed out the lack of specific details of what was meant by the proposal. Others have noted that such a focus creates yet another responsibility for third world women and again marginalises debate on the global and national maldistribution of resource consumption.

Both Rio and Cairo illustrate that the benefits of such conferences are difficult to measure. Apart from the 'soft' agreements and the brief but intense focus of the international media, delegates (both governmental and non-governmental) established networks of contacts that live on after the event. In addition, each conference manages to take on board major criticisms of the previous event. For example, 'Habitat 2' at Istanbul in 1996 did manage to address the impact of trade liberalisation on poorer communities, particularly with reference to issues of shelter.

International environmental politics, as practised in the series of United Nations-inspired conferences (like Rio and Cairo), has been both effectual and ineffectual. Such conferences are good for producing formal statements, but they have failed to establish international institutions to monitor and regulate global environments. On the other hand, they have been extremely effective in promoting the meta-narrative of sustainable development (with its advocacy of the globalisation of market systems and pluralist systems of democracy) as the only way out of 'the global environmental crisis' (Doyle 1996). In a book detailing the events that took place at Cairo, Stanley Johnson defends the conference in this vein:

> With the full 'empowerment of women', as proposed in the Cairo programme, the goals of sustainable development will be that much easier to achieve, and the goal of population stabilisation will no longer be a distant dream but a real, practical possibility.
>
> (1995: 10)

'Ozone diplomacy'

The fate of the ozone layer in the upper atmosphere graphically illustrates one of the most significant dimensions of the global challenge of environmental problems (Benedict 1989; Pearce 1991: ch. 5). It illustrates some important characteristics of an environmental issue that can slip beyond a response by governments focused on their own countries and locked into a system premised on national sovereignty. It is also a tale that can illustrate the conditions under which global problems like this can be addressed.

The problem of ozone depletion starts with the discovery of the industrial properties of a series of new chemicals exemplified by the chlorofluorocarbons (CFCs). These have amazing properties of expansion and compression, which made them ideal as propellants for spray cans and for the cooling systems in refrigeration or air-conditioning units. In the short run, there was no obvious environmental problem, as the gases appeared to be inert under normal conditions. It was only later (but not all that much later, since the major increase in use of CFCs came in the 1950s and 1960s) that a change in the upper atmosphere was noted. In popular language, this became known as the 'hole' in the ozone layer, initially identified over Antarctica but now also apparent over the Arctic Circle. What happened was a simple and predictable chemical reaction. When CFCs escape from cooling systems or industrial processes, they gradually find their way into the upper atmosphere, where they break down and react with ozone. The result is a reduction in the natural level of ozone in the upper atmosphere. This might have been unimportant but for the fact that ozone works as a filter against the penetration of ultraviolet (UV) radiation from the sun. As the ozone layer thins, greater quantities of UV radiation get through to the Earth's surface. This extra UV can affect human and animal health (at the very least, increasing the incidence of skin cancer) and can damage plant tissues (which also reduces crop yields).

As an environmental problem, this has all the characteristics of a fully global issue. Although the USA was the greatest producer of ozone-depleting substances, no single country could, by its own actions, solve the problem. For example, Australia, being in the southern hemisphere, may detect an increase in the incidence of skin cancer associated with the thinning of the ozone layer, but no policy initiatives taken by Australia acting on its own can stop the damage continuing. Further, countries that produce no CFCs and use little are just as susceptible to the consequences as those that produce a lot. In addition, the structure of the global economy and increasing industrial development means that there is scope for greater production and use of CFCs, especially as China industrialises. Stopping the use of CFCs would have differential economic effects depending on the degree of development. In all these circumstances, there was every chance of policy stalemate, inaction and increasing damage to the upper atmosphere.

In this case, ways were found to build a co-operative international (not quite global) response to this environmental problem. In essence, the solution was built on diplomacy and the negotiation of a treaty regulating CFCs, including banning their use for some purposes. It was hoped that this regime of international regulation would reduce CFC emissions and in time, allow the ozone layer to recover.

Stated in this way, there is no indication of the special conditions that were required to produce international co-operation over a global issue when this is more an exception than a norm. Two factors intersected to make this treaty approach viable. First, there was the chemistry and commercial possibilities of a CFC replacement. Research chemists working for the big chemical companies began an early search for a replacement, and the efforts by those companies to delay regulation (by lobbying and stressing the uncertainty of the science surrounding the ozone issue) provided time to develop a commercially viable alternative. Then the efforts to ban CFCs could be turned to commercial advantage by those companies with the substitute ready to go into production. This breakthrough came first in the USA and, at an opportune time, was developed by a leading German firm so that the interests of chemical companies could begin to work through negotiations between rival nation-states. This commercial development of an alternative to CFCs is important because it meant that the impact of the ban on economic growth was both slight and advantageous to a couple of significant firms.

Second, there is the role of the USA in building and policing an effective international environmental regime. If the United States was opposed to regulation, if US-based chemical companies were to be disadvantaged by any restriction on production and use, then the chances of international co-operation would have been slight. Indeed, initially, both US companies and the government opposed international regulation, but their position changed and the US government assumed leadership of the diplomatic negotiations that culminated in the Montreal Protocol on Substances that Deplete the Ozone Layer, adopted in 1987. Following the changed position of the USA, West Germany also changed its perspective and the leadership it provided inside what was then the European Economic Community.

On the back of these negotiations, a regime of international regulation was built and there have been reductions in the use of CFCs and a slowing in the rate of ozone depletion. But it is important to note that this regime is not fully effective. CFCs are still produced and used in countries that are signatories to the protocol; the substitute itself is capable of adding to ozone depletion but at a vastly slower rate; and industrialising countries are using CFCs in their refrigeration technology and the transfer of technology has been slow, costly and insufficiently financed by the development funds associated with international regulation. Thus, although the international approach to ozone depletion is a significant example of how international co-operation can be constructed to respond to global environmental problems, it also reveals the fragility and limits of such an approach.

The international politics of climate change

The issue of climate change is like that of ozone depletion in that it concerns a global problem that transcends national boundaries and is beyond the scope of any single nation to reverse the problem (Pearce 1991: chs 3, 4). But it is also unlike the ozone problem in other ways, and considering these differences identifies factors that stand as barriers to effective resolution. The problems involved in responding to the prospect of global warming illustrate, in more depth, the difficulties faced in building effective international environmental regimes.

The issue of international climate change exemplifies all the problems of scientific modelling of major system-wide changes. The problem can be simply described, at least as a hypothesis and then as a prediction of likely developments. Industrialisation has been based on the use of fossil fuels to produce energy. When burned, these fuels release carbon dioxide (CO_2) into the atmosphere. As energy consumption has increased so have concentrations of CO_2 in the atmosphere. More atmospheric CO_2 increases the amount of long-wave radiation trapped by the Earth's atmosphere and enhances the natural greenhouse effect, which keeps the biosphere within a temperature range that is able to support life. On the back of evidence of increasing concentrations of carbon dioxide in the atmosphere, complex computer models of the global weather system and arguments about an enhanced greenhouse effect come predictions of global warming, which can change the world's climate patterns and produce major changes in the world's environment. The world's mean temperature can rise, sea levels increase, deserts expand and climate bands shift, with major consequences for plants and animals.

It is important to understand some of the underlying complexity of the science and modelling involved in the prediction of global warming. Global temperatures fluctuated long before fossil fuel use increased, with periods of warming and cooling following 'natural' causes. Records of global warming or cooling are difficult to establish and are incomplete. Other factors complicate the position: natural processes like volcanic eruptions can increase the presence of 'greenhouse gases' in the atmosphere, but they also throw out particulate matter that has a cooling effect because of its ability to deflect part of the Sun's heat; increased air pollution and greater cloud cover (from the increased evaporation caused by higher temperatures) can also produce this cooling effect; and finally, the impact of the world oceans as heat reservoirs and CO_2 sinks is unclear. The hypothesis of the enhanced greenhouse effect is fairly simple, but the prediction of its consequences for global weather is based on two sets of complex computer modelling. The first deals with how the global climate systems work. The second models the processes that could link increased CO_2 concentrations to increased warming. In all such modelling exercises, the whole sequence of equations that make up the model are based on assumptions, hypotheses and odd bits of 'hard' information. Such uncertainty always makes the debate on the greenhouse effect complex and open to a variety of interpretations. Global warming is an environmental issue defined by its complexity, the ambiguity in the analysis and the long time scale for both effects and counter-measures.

The barriers to an effective international response to the prospects of an enhanced greenhouse effect are far greater than those that operated over ozone depletion. Despite their utility to industry, CFCs are not the foundation for an effectively operating national economy. Energy production and consumption are. Britain, the USA, Japan and Germany all industrialised on the back of relatively cheap energy and only limited pressures to deal with its polluting consequences. Pressure to reduce greenhouse gas emissions can therefore exert quite significant pressures on the rate of economic growth and the scope for industrialising countries to follow on the path of the now-dominant economies. Hence both industrialised and industrialising countries can share the economic benefits of not responding to the greenhouse challenge. Significant private interests in the form of coal and oil producers and electricity companies stand to lose significantly from serious efforts to reduce CO_2 emissions. Unlike the ozone case, there are no ready, cheap alternatives to substitute for fossil fuels.

As a reflection of this complexity and the strong forces arrayed against action and the scope for coalition building to ignore the greenhouse challenge, there has been only a limited international response. Global warming has been debated in numerous global forums and resolutions calling for (voluntary) action have been endorsed. Targets for reduction have been set and reporting procedures have been introduced, but very little has been achieved.

The Australian case is, in this instance, illustrative. The Australian government made the greenhouse effect one of the topics to be debated in the Ecologically Sustainable Development process. Similarly, the economic consequences of greenhouse gas reduction strategies were referred to the Industry Commission, a body charged with making policy recommendations to improve the competitive efficiency of the Australian economy (essentially by recommending an end to various forms of government subsidies, regulation and other forms of intervention). The government adopted a greenhouse strategy based on predictions that taking serious action would reduce economic growth and that the benefits were uncertain. It was also mindful of the opposition of the powerful resource and energy companies and the significant part that coal exports played in Australia's balance of trade. Hence modest reduction targets were set and the emphasis was on voluntary action in a 'no regrets' framework. These were, of course, inadequate for the task of meeting previously agreed targets. As most countries failed to meet their modest international commitments, international pressure grew for legislated and mandated responses. This was bolstered by the demands from small, generally insignificant players on the international scene like the Pacific island nations, which would be threatened by even modest rises in sea levels due to global warming; by increased environmental concern; and by the increasing scientific consensus behind the global warming hypothesis. Australia has tried very hard in international forums to prevent mandated targets from being approved. It must have been an educative experience to see the Australian environment minister attending such forums bolstered by the presence and support of Australia's energy producers arguing against greater effort on greenhouse gas reductions. Under President Bush, the United States, a major energy producer and exporter, initially opposed stricter measures for the Rio Convention on Climate Change. Since Clinton's election, Australia's position has been increasingly isolated, with the USA willing to call for greater and more mandated action.

In general, action on greenhouse gas reduction has been less determined, less agreed and less effective than action on ozone depletion. The chances of successful action are slight even if treaties are signed and enforced. The desperate need for economic development in large parts of the globe, the unwillingness of advanced industrial powers to share technology and the fruits of growth, the competitive struggle between these industrial powers, and the significant economic consequences of determined action suggest that this will not be one of the success stories of international environmental regime building.

Conclusion

The basic lessons of these examples reveal the limitations inherent in a system of nation-states responding to environmental problems that are shared, problems that cross national boundaries and those that are truly global. There is plenty of scope for global forums where world leaders meet and NGOs gather to discuss shared problems and exchange views on possible policy instruments and experience. Such forums are the focus for the preparation of major statements, such as *Agenda 21*, *Our Common Future* and the occasional protocol. There is much talk, much paper is consumed, many media events are staged and sometimes global awareness is increased, but these are not usually the point at which things get done. Global forums and diplomacy can produce solutions or the frameworks for addressing some environmental problems, such as ozone depletion, but there is no simple blueprint for success. Some problems, like global warming, test the limits of the ability to respond. The most enduring challenge for environmental politics and policy making lies at this international level.

Further reading

Chatterjee, P. and Finger, M. (1994) *The Earth Brokers: Power, Politics and World Development*, Routledge, London and New York.

Pepper, D. (1984) *The Roots of Modern Environmentalism*, Croom Helm, London.

Pepper, D. (1993) *Eco-Socialism: From Deep Ecology to Social Justice*, Routledge, London and New York.

Conclusion: environment and politics

In this book, we have provided an introduction to understanding the comparative dimension of environmental policy and policy making. Environmental problems are either shared between, or common to, a large number of countries, so it is possible to consider the similarities and differences between their responses. To make an effective comparison it is necessary to describe what is being done in response to environmental challenges and to note variations between countries and over time. It is also necessary to have a framework within which the comparison can be evaluated. In this book, we have concentrated on providing a series of alternative ways of analysing and evaluating what happens in different countries and at different times. We have been concerned to give accounts of the different concepts that are necessary for the comparative enterprise, as well as the different arguments that are made both about these concepts and the ways in which these interpret responses to environmental problems. In addition, we have illustrated our presentation with examples drawn from a wide range of countries in quite different situations. For instance, we have drawn examples from the United States of America, Britain, Western Europe and Australia to illustrate the response in relatively wealthy, well-industrialised countries with liberal-democratic political systems. Contrasts have been sought from the countries of Eastern Europe, which experienced environmental degradation as a result of state action pursuing rapid industrialisation, and from countries like South Africa, which are caught between an urgent need for economic development and concern about its environmental consequences.

Central to the presentation of the opportunity for a comparative assessment of environmental politics is a concern with its necessary interdisciplinary character, bringing together in a single frame the contributions of a wide range of disciplines of both the physical and social sciences. A series of core concepts have been discussed. The character of the political system, from liberal-democratic through to authoritarian rule, has an important impact on what is done (or not done) in the name of environmental concern. The concept of power, from behaviourists and pluralists through to the neo-pluralists and Foucauldians, is central to an understanding of what is done, why it is done and how these actions are to be interpreted.

From a consideration of the core concepts of political regimes and power, the discussion moved on to an account of the wellsprings of environmental action: the range of arguments, claims and counter-claims that impel different responses to environmental problems. For example, some claim that nothing needs to be done, since environmental problems are either not serious or are easily solved. Others note environmental problems and propose reforms but seek to act within the prevailing assumptions of economic growth. There are still others who have developed more radical accounts of what causes environmental damage and what cure is needed. This last group has formed the basis for the most successful attempts to move 'green' political concerns into the general political process. Having established the grounds for political action based on environmental concerns, it is necessary to consider the form that political action takes, from social movements and non-governmental organisations through to fully independent green political parties.

Environmental politics is not only made by those who are motivated by environmental concern. Even those who generally resist such concerns might find themselves seeking to accommodate them in particular areas. Given the significance of private business to economic growth, its attitude to environmental damage and its response is of great importance. Further, what happens as a result of environmental politics (frequently the product of conflict between those promoting economic development and those promoting environmental concern) – the making of policy – is heavily conditioned by the actions of the bureaucracy and of administrative practice. There are several different ways of making and implementing policy and a variety of ways to evaluate the consequences of administrative responses to environmental problems and policy making.

There is a sense in which comparative study reproduces in analysis the assumptions of national sovereignty. For many political issues this is not important, but for environmental issues it is. Where those issues are either shared or common, where they cross national boundaries or where they are truly global, then what goes on inside the nation-state is not a sufficient focus. Here it is necessary to consider the extent to which international politics and the international system is able to deal with the complexity of global environmental concerns. This necessarily complements the arguments made about understanding the comparative dimension of environmental politics.

Overall, this book has set out the analytical tools that can be used to craft understandings of the sources and character of domestic responses to both local and global environmental problems, as well as the limitations of this response. The same could be said of the international response. The world is confronted by a whole series of environmental problems, especially focused on how to improve the conditions of life through economic growth, while limiting the damage that comes from economic development. In considering the different options that can be applied, the comparative perspective is very useful: it gives a disciplined approach to considering what gives strength or weakness to different systems and different ways of responding to the environmental challenge.

⬤ Notes

Introduction

1 'North' and 'South' are crude terms that try to name a complex, differentiated and nuanced reality. There are no simple, agreed, ambiguous and effective terms to describe the distribution of power and affluence in the global order. The terms 'North' and 'South' were at their most important back in the days of the Brandt Report and the UNCTAD conferences, where they tried to capture in a geographic/spatial metaphor the maldistribution of power and economic strength between the United States of America, the countries of Western Europe and Japan, in contrast with poorer countries then seeking to industrialise. The spatial metaphor fails as it lumps together the rich and poor within these countries and suggests that they share common perspectives, interests and concerns. Even adding the qualifying 'elites' and 'broad population' does little to make the nomenclature more effective. Parallel terms developed/undeveloped or first world/third world share similar difficulties. Although we use these terms as a piece of convenient shorthand we are ever mindful of the problems they cause for analysis if they are not treated with extreme caution.

2 Martin Seliger defines ideology in two ways: restrictive and inclusive (Seliger 1976). Restrictive ideology necessitates a dominant group in society which constructs a system of beliefs and values to further its own, elite goals. Ideology, in this sense, is used to repress the majority of people residing outside this dominant grouping. Inclusive models portray ideology as a multitude of different sets of beliefs, ideas, myths, values, etc. belonging to separate groups in all societies. In this latter definition, all people have an ideology. Obviously, all people do have an ideology – a view of the world – but it must be recognised that some ideologies are more powerful and persuasive than others.

Chapter 1

1 Some of this unreality was picked up by pluralists in their subsequent analytical careers. Charles Lindblom, who was one of the seminal figures in the development of pluralist accounts, went on to write a classic account of the consequences of the preponderance of power held by business in the US political system, *Politics and Markets* (Basic Books, New York, 1977). Dahl himself went on to write a number of works that were concerned

with the unequal distribution of economic power and its consequences for a democratic polity; see *After the Revolution* and *Economic Democracy*.

2 The concept of 'interests' is one of the most contested in methodological debates of political and social science. There are as many issues to be raised about the conception and study of interests as there are about the study of power. Some of the elements are covered by Lukes (ch.6) but see also Connolly (1972) and McEachern (1980).

3 In the vast array of Foucault's work, it is not always clear that there is a unified, consistent 'Foucauldian' interpretation of power, although there are many attempts to state one. The best account is provided by Hindess (1996), who both notes the diversity of Foucault's positions and an important change in his conception of power at about the time he gave his lectures on governmentality and published the *History of Sexuality*, Vol. I.

Chapter 2

1 This is not true of the most important of the free market environmentalists, David Pearce, whose work is an exemplary effort at finding market solutions to environmental problems that are treated seriously. Further, David Pearce has sought more than most to take sustainability into the very construction of the basic theorems of neo-classical economics. A most accessible version of the arguments developed by Pearce and his associates is to be found in Cairncross (1991).

2 A joint publication of the International Union for Conservation of Nature and Natural Resources (IUCN), the United Nations Environment Programme (UNEP) and the World Wildlife Fund (now the World-Wide Fund for Nature) (WWF).

3 World Commission on Environment and Development (WCED) chaired by Gro Harlem Brundtland. Australian edition published with additions from the Commission for the Future in 1990.

4 Only some types of eco-feminism can be regarded using this oppositional, paradigmatic model. For example, many parts of eco-feminism are closely related to liberal sensibilities and, as such, are more concerned with seeking eco-feminist change through accommodation within existing political systems (as investigated in the previous section).

5 Warrick Fox lists four: ethical sentientism, biological ethics, ecosystem ethics (Gaian) and cosmic purpose ethics.

Chapter 3

1 Post-materialism and post-industrialism are not mutually exclusive and share several propositions.

2 This argument about the palimpsest is presented in greater detail in Doyle, T.J. and Kellow, A.J. (1995) *Environmental Politics and Policy Making in Australia*, Macmillan, Melbourne.

3 See Chapters 4 and 8 of Doyle, T.J. and Kellow, A.J. (1995) *Environmental Politics*

and Policy Making in Australia, Macmillan, Melbourne, for a close analysis of the politics of informal environmental groups and associations operating in liberal democracies such as Australia.

4 In the United States in the early 1990s, 95 percent of professional staff and volunteer leaders from over 500 conservation and environment groups nationwide agreed with the following statement: 'Many, perhaps most, minority and poor rural Americans see little in the conservation message that speaks to them.' Not one major environment organisation could boast significant black, Hispanic or Native American membership. See C. Jordan and D. Snow, 'Diversification, Minorities, and the Mainstream Environmental Movement', in Snow, D. (ed.) (1992) *Voices From the Environmental Movement: Perspectives for a New Era*, Island Press, Washington, DC.

5 As aforesaid, there are numerous exceptions. There are active wilderness-oriented movements, for example, in parts of Africa.

6 In 1990, the United Nations Human Development Report read: 'poverty is one of the greatest threats to the environment.' In 1993, an International Monetary Fund article read: 'Poverty and the environment are linked in that the poor are more likely to resort to activities that can degrade the environment' (Broad 1994: 812).

7 For an excellent example of this line of reasoning read Hartshorn, G.S. (1991) 'Key Environmental Issues for Developing Countries', *Journal of International Affairs*, vol. 3: 393–401.

8 In many well-meaning works prior to the Rio Summit in 1992, it was constantly argued that the South had to become more like the United States in its political system if environmental degradation was to be brought under control. In this vein, the development of 'civil society' is central to the discussion. Underlying this 'civil society' is an unquestioned acceptance of the American form of democracy based on capitalism and global free markets. An excellent example is Ghai, D. and Vivian, J. (eds) (1992) *Grassroots Environmental Action: People's Participation in Sustainable Development*, Routledge, London, New York. They write: 'The existence of a democratic space allowing the expression and defence of community rights and claims has proven to be a crucial factor influencing successful grass-roots environmental action. . . . The essence of these activities is to persuade or pressure the state to intervene on behalf of the communities through adoption of new legislation.' (18–19).

9 In an article by Robin Broad on the Philippines' environmental movement, 'The Poor and the Environment: Friends or Foe', *World Development*, vol. 22, no. 6., 812–22, 1994, an important distinction is made on the connections between *types* of poverty and environmental degradation. Broad distinguishes between the 'merely poor' and the 'very, very poor'. Fundamentally, the former category are those still operating subsistence lifestyles (though under threat) and those who have recently been removed from this lifestyle. The latter category are the 'landless and rootless'. These have no security of tenure and little connectedness to place. This category includes those peasants and squatters who survive by cutting forest cover, by consuming wildlife, and by planting crops on soils that will erode.

Chapter 4

1 See Carol Deal's booklet prepared for Greenpeace entitled *The Greenpeace Guide to Anti-Environmental Organizations*, Odonian Press, Berkeley, California, 1993. The work looks at six types of 'anti-environmental' organisations, including public relations firms, corporate front groups, think-tanks, legal foundations, endowments, charities, and wise use and share groups. One excellent example of a corporate front group posing as a green NGO is the British Columbia Forest Alliance in Canada. The Canadian timber industry paid Burson-Marstellar one million dollars to create the alliance. Deal writes: 'Like its US counterparts – the Evergreen Foundation and the National Wetlands Coalition – the alliance has two tasks: convincing the public that the current rate of environmental destruction can be maintained or increased without long-term effect, and persuading lawmakers to roll back unprofitable environmental regulations' (16–17).

2 One study performed during *glasnost* counted 331 environmental organisations in the Russian Federation and the Ukraine (Princen and Finger 1994: 2). Almost no environmental organisations existed in the former Soviet Union before this date. In Indonesia, still playing its politics under an authoritarian regime, environmental NGOs have increased from 79 in 1980 to over 500 in 1992.

3 Earth First! is as non-instituional as organisations get. Some Earth First! participants refer to it as a non-organisation. In fact, it does have many of the attributes of informal groups described in the previous chapter. It falls on the cusp between these two collective forms. But the goals of Earth First! are often listed in its magazine *Earth First!* These goals seem relatively continuous and serve some of the purposes of a constitution.

4 Chatterjee and Finger write of these direct, sometimes militant, eco-actions: 'Yet their tactics have often been surprisingly effective. The Sea Shepherds closed down the Icelandic whaling industry singlehandedly one cold November night in 1986 by the simple expedient of sinking two of its four ships and destroying the refrigeration system of its whale processing plant. . . . Ecosaboteurs in Canada blew up a US$4.5 million hydro-electric substation on Vancouver Island in 1982. In Thailand they burnt down a tantalum plant in 1986 causing damage estimated at US$45 million. Lapps in Norway blew up a bridge leading to a dam that had flooded their lands. Then further out from even the deep ecologists are people who do accept physical injury or death as punishment. . . . For example, Primea Linea, an Italian group, claimed responsibility for machine gunning Enrio Paoletti, an executive of a Hoffmann-LaRoche subsidiary, who was in charge of the chemical plant in Seveso, Italy, that exploded in 1976 to release a dioxin cloud' (1994: 72).

5 These private purchases of land are common. In 1996, the US Campaign to Save Mount Jumbo in Missoula, Montana, raised sufficient money to buy the elk habitat outright.

6 In 1995, Timothy Doyle interviewed participants enrolled in the United Nations Environment Programme International Certificate in Environmental Management, at the University of Adelaide. The majority of participants worked in the middle and upper management positions in their own country's equivalent of 'environment departments'. All participants were from nations of the 'developing world'. Information derived from this written questionnaire is included in this section. Full results available from the author.

7 For a detailed account of internal politics in Greenpeace, FoE, the ACF and the

Wilderness Society see Doyle, T. J. (1998) *Green Power: The Politics of the Environment Movement in Australia*, Scribe Publications, Australia (in press).

8 There are huge regional differences in the operation of FoE. In Australia, for example, there is much autonomy within the organisation, whereas the US experience is far more centralised and hierarchical.

9 See Timothy Doyle, 'Oligarchy in the Conservation Movement: Iron Law or Aluminium Tendency?', *Regional Journal of Social Issues*, Summer 1989, 28–47. This article documented an internal power struggle within two environmental organisations in Australia, the ACF and the Wilderness Society, in the late 1980s. Elite theories were drawn upon to provide a theoretical discussion.

Chapter 5

1 There have been minor successes in Eastern Europe. For example, the Green Party in Hungary received 3.7 percent of the vote at the national level in the elections of spring 1990. This was not enough, however, to secure representation in the parliament (Szabo 1994: 294). Since the 'democratic transition' green parties have formed in Eastern Europe but have not been significant players. Interestingly, several green parties have emerged in the outlying states of the former Soviet Union (Richardson and Rootes 1995: 20).

2 These points were based on empirical studies of the anti-uranium movement and the movement against the flooding of the Franklin River in southwest Tasmania, Australia. Both these movements were extremely active in the late 1970s and 1980s.

3 We accept that there are contextual differences in the terms 'environmental', 'ecological' and 'green'. But, as mentioned at the outset, for the purposes of establishing a working definition for this book, we use the terms interchangeably.

4 At the local level, a Green councillor, John Gormley, became Lord Mayor of Dublin in early 1994 (Holmes and Kenny 1994: 222).

5 Obviously, there are also many differences between the US and British models. Additionally, within each country there are many different types of election (sometimes working on separate models), ranging from the local level, through the national, to the European Union elections, in the case of Britain; and from local school boards, through the state and federal congresses, to the presidency, in the instance of the USA. The US presidential election is an excellent example of the first-past-the-post system. Each state in the federation has a distinct number of 'electoral college' votes. If one candidate wins the most votes in the state, then he/she takes the entire number of electoral college votes from that state and adds them to his/her national tally. This occurs regardless of the fact that he/she may have the support of only one more *elector* than another candidate in the state in question. For an introductory guide to this process see O'Connor, K. and Sabato, L. J. (1993) *American Government: Roots and Reform*, second edition, Allyn and Bacon, Boston: 515–27.

6 All states use this system except Tasmania, which utilises the complicated Hare–Clark system. This system is fundamentally one of proportional representation, and it partly accounts for the unusually good showing of Tasmanian greens in this state's electoral polity.

Chapter 6

1 Gro Harlem Brundtland chaired the UN World Commission on Environment and Development (WCED), which operated from 1983 to 1987 and produced *Our Common Future*.
2 The Rio Earth Summit's formal title was the United Nations Conference on Environment and Development (UNCED). The major communiqué from the conference was *Agenda 21*.

Chapter 7

1 These should be contrasted with David Pearce's unrelated efforts to add a theorem of sustainabilty to the foundational axioms of neo-classical economics. See Pearce (1991) for the working through of this project and its radical consequences for assessing policy options. In effect, Pearce tries to add a concept of ecological rationality to a field of economic rationality, where rationally linking means and ends is complicated by treating sustainability as if it mattered.
2 For a clear and systematic discussion of this point see Hindess, *Discourses of Power*.
3 The broad shape of these measures has been assessed in Chapter 6, and the comments on those who oppose environmental concern in the name of economic efficiency are not repeated here.
4 Environmentalists from the ACF, the Wilderness Society and Greenpeace initially participated in eight of the nine ESD working groups. Greenpeace withdrew from the process because of proposed Commonwealth resource security legislation. All groups refused to sit on the Forest Industry group from the beginning and pursued their case through RAC instead.

References

Aditjondro, G. and Kowalewski, D. (1994) 'Damning the Dams in Indonesia: A Test of Competing Perspectives', *Asian Survey*, XXXIV, 4: 381–95.

Annis, S. (1992) 'Evolving Connectedneess Among Environmental Groups and Grassroots Organizations in Protected Areas of Central America', *World Development*, 20, 4: 587–95.

Arendt, H. (1970), *On Violence*, Allen Lane, London.

Australian, The, 6 September 1996.

Bachrach, Peter and Baratz, Morton S. (1962) 'The Two Faces of Power', *American Political Science Review*, 56, 947–52.

—— (1963) 'Decisions and Nondecisions: An Analytical Framework', *American Political Science Review*, 57, 641–51.

—— (1970) *Power and Poverty. Theory and Practice*, Oxford University Press, New York.

Baker, S., Milton, K. and Yearly, S. (eds) (1994) *Protecting the Periphery: Environmental Policy in Peripheral Regions of the European Union*, Frank Cass & Co., Essex.

Bartlett, Robert V. (1990) 'Ecological Reason in Administration: Environmental Impact Assessment and Administration Theory', in Robert Paehlke and Douglas Torgerson (eds) *Managing Leviathan: Environmental Politics and the Administrative State*, Belhaven Press, London, 1990: 81–96.

Bean, C. and Kelley, J. (1995) 'The Electoral Impact of New Politics Issues: The Environment in the 1990 Australian Federal Election', in *Comparative Politics*, April: 339–56.

Bebbington, A. and Thiele, G. with Davies, P., Prager, M. and Riveros, H. (1993) *Non-Governmental Organizations and the State in Latin America: Rethinking Roles in Sustainable Agricultural Development*, Routledge, London and New York.

Beckerman, Wilfred (1995) *Small is Stupid: Blowing the Whistle on the Greens*, Duckworth, London.

—— (1974) *In Defence of Economic Growth*, Jonathan Cape, London.

—— (1990) *Pricing For Pollution* (second edition), Institute for Economic Affairs, London.

Bell, S. (1987) 'Socialism and Ecology: Will Ever the Twain Meet', *Social Alternatives*, 6, 3: 5–12.

Benedict, R. (1991) *Ozone Diplomacy: New Directions in Safeguarding the Planet*, Harvard University Press, Cambridge, Mass.

Bennett, Jeff and Block, Walter (1991) *Reconciling Economics and the Environment*, West Perth, Australian Institute for Public Policy.

Boggs, C. (1986) *Social Movements and Political Power: Emerging Forms of Radicalism in the West*, Temple University Press, Philadelphia.

Bookchin, M. (1980) 'Ecology and Revolutionary Thought', in R. Roelofs, J. Crowley and D. Hardest (eds) *Environment and Society*, Prentice Hall, New Jersey.

—— (1987) 'Social Ecology versus "Deep Ecology": A Challenge for the Ecology Movement', *The Raven Anarchist Quarterly*, 1, 3: 219–50.

Bookchin, M. and Foreman D. (1991) *Defending the Earth: A Dialogue between Murray Bookchin and Dave Foreman*, South End Press, Boston.

Boreham, G. (1996) 'Keating Woos Greens', *The Age*, 25 January.

Bramwell, A. (1994) *The Fading of the Greens: The Decline of Environmental Politics in the West*, Yale University Press, New Haven, Conn., and London.

Broad, R. (1994) 'The Poor and the Environment: Friends or Foes?', in *World Development*, 22, 6: 811–22.

Bullard, R.D. (ed.) (1993) *Confronting Environmental Racism: Voices from the Grassroots*, South End Press, Boston.

Bureau of Industry Economics (1992) *Environmental Regulation: The Economics of Tradeable Permits – A Survey of Theory and Practice*, Research Report 42, AGPS, Canberra.

Cairncross, Francis (1991) *Costing the Earth*, Economist Books, London.

Callenbach, E., *et al.* (1993) *EcoManagement: The Elmwood Guide to Ecological Auditing and Sustainable Business*, Berrett-Koehler, San Francisco.

Carlassare, E. (1994) 'Essentialism in Ecofeminist Discourse', in C. Merchant (ed.) *Ecology: Key Concepts in Critical Theory*, Humanities Press, New Jersey.

Carter, F. W. and Turnock, D. (1993) *Environmental Problems in Eastern Europe*, Routledge, New York.

Chaloupka, B. (1996) 'The Year of the Green', *Missoula Independent*, March 28: 17.

Chatterjee, P. and Finger, M. (1994) *The Earth Brokers: Power, Politics and World Development*, Routledge, London and New York.

Cheney, J. (1989) 'Postmodern Environmental Ethics: Ethics as Bioregional Narrative', *Environmental Ethics*, 11: 117–34.

Cock, J. and Koch, E. (eds) (1991) *Going Green: People, Politics and the Environment in South Africa*, Oxford University Press, Cape Town.

Connolly, William E. (1972) 'On "Interests" in Politics', *Politics and Society*, Vol. 2, 459–77.

Cooper, M. (1994) 'The Greens Climb in New Mexico', *The Nation*, October, 259, 3: 453–7.

Coyote, H. (1991) 'The Corporate Takeover of Friends of the Earth', *Chain Reaction*, 63/64, April: 35–8.

Crenson, Matthew A. (1971) *The Unpolitics of Air Pollution: A Study of Non-Decision Making in the Cities*, Johns Hopkins University Press, Baltimore, Md.

Crick, B. (1964) 'The Nature of Political Rule', *In Defence of Politics*, Allen Lane, London.

Cuomo, C. (1992) 'Unravelling the Problems in Ecofeminism', *Environmental Ethics*, winter edn: 351–63.

Curtis, Bruce (1995) 'Taking the State Back Out: Rose and Miller on Political Power', *British Journal of Sociology*, Vol. 46, No. 4, 575–89.

Cushman, John H. (1996) 'Adversaries Back the Current Rules Curbing Pollution,' *New York Times*, (Monday, February 12), pp. 1 & C11.

Dahl, Robert A. (1957) 'The Concept of Power', *Behavioural Science*.

—— (1961) *Who governs? Democracy and Power in an American City*, Yale University Press, New Haven, Conn., and London.

—— (1970) *Modern Political Analysis*, Prentice Hall, New Jersey.

Dalton, R.J. (1994) *The Green Rainbow: Environmental Groups in Western Europe*, Yale University Press, New Haven, Conn., and London.

Deal, C. (1993) *The Greenpeace Guide to Anti-Environmental Organizations*, Odonian Press, Berkeley, Calif.

Dean, Mitchel (1991) *The Constitution of Poverty*, Routledge, London.

Department of Home Affairs and Environment (1982) *A National Conservation Strategy for Australia*, AGPS, Canberra.

De Shalit, A. and Talias, M. (1994) 'Green or Blue or White? Environmental Controversies in Israel', *Environmental Politics*, 3, 2: 273–94.

Detlef, J. (1994) 'Unifying the Greens in a United Germany', *Environmental Politics*, Summer, 3, 2: 312–18.

Devall, B. and Sessions, G. (1985) *Deep Ecology*, Peregrine Smith Books, Salt Lake City, Utah.

DiLorenzo, Thomas J. (1993) 'The Mirage of Sustainable Development', *The Futurist* (September–October), 14–19.

Ditz, D., Ranganathan, J. and Banks, D. (1995) *Green Ledgers: Case Studies in Corporate Environmental Accounting*, World Resources Institute, Washington DC.

Dobson, A. (1995) *Green Political Thought* (second edition), Routledge, London and New York.

Douglas, M., Lee, Y.S.F. and Lowry, K. (1994) 'Introduction to the Special Issue on Community Based Urban Environmental Management in Asia', *Asian Journal of Environmental Management*, 2, 1: ix–xv.

Doherty, B. (1992) 'The Fundi-Realo Controversy: An Analysis of Four European Green Parties', *Environmental Politics*, 1, 1: 95–120.

Dowdeswell, E. (1994) Speaking Notes for Elizabeth Dowdswell, Under-Secretary General and Executive Director United Nations Environment Programme, electronic preparation by the Population Information Network of the United Nations Population Division in collaboration with the United Nations Development Programme, September 6.

Dowie, M. (1994) 'The Selling (Out) of the Greens', *The Nation*, April 18: 514–18.

—— (1995) *Losing Ground: American Environmentalism at the Close of the Twentieth Century*, The MIT Press, Cambridge, Mass., and London.

—— (1995) 'The Fourth Wave: An Opinion by Mark Dowie', *Mother Jones*, March/April: 34–6.

Doyle, Timothy (1986) 'The "Structure" of the Conservation Movement in Queensland', *Social Alternatives*, Vol. 5, No. 2: 27–32.

—— (1989) 'The Conservation Movement and the Aluminium Tendency of Oligarchy', *Regional Journal of Social Issues*, Summer issue, 28–47.

—— (1991) 'Informal Groups and the Conservation Movement: A Matter of Introspection', Conference of the Australian Political Studies Association, Griffith University, July 17–19.

—— (1994a) 'Direct Action in Environmental Conflict in Australia: A Re-examination of Non-Violent Action', *Regional Journal of Social Issues*, Australia, 28: 1–13.

—— (1994b) 'Dissent within the Environment Movement', *Social Alternatives*, Vol. 13, No. 2: 24–6.

—— (1996) 'Agenda 21: Ecological Imperialism and the Globalisation of Environmental Management' *Ecopolitics X*, Proceedings of the Conference, Australian National University, September.

—— (1997) *Green Power: The Environment Movement in Australia*, Scribe Publications, Melbourne (in press).

Doyle, Timothy and Kellow, Aynsley (1995) *Environmental Politics and Policy Making in Australia*, Macmillan, Melbourne.

Doyle, Timothy and Walker, Ken (1996) 'Looking for a World They Can Call Their Own', *Campus Review*, Australia, February 1–7: 10, 16.

Dryzek, John S. (1987) *Rational Ecology: Environment and Political Economy*, Basil Blackwell, Oxford.

—— (1990) 'Design for Environmental Discourse: The Greening of the Administrative State?', in Robert Paehlke and Douglas Torgerson (eds) *Managing Leviathan: Environmental Politics and the Administrative State*, Belhaven Press, London, 97–111.

Duverger, M. (1972) *The Study of Politics*, Crowell, New York.

Earth Times Foundation (1996) *The Earth Times*, IX, 3, Geneva and New York.

Easterbrook, G. (1995) *A Moment on the Earth: The Coming of Age of Environmental Optimism*, Viking Penguin, New York.

Eckersley, R. (1990) 'The Ecocentric Perspective', in C. Prybus and R. Flanagan (eds) *The Rest of the World is Watching*, Pan Macmillan, Sydney.

Eckersley, R. (1992) *Environmentalism and Political Theory: Toward an Ecocentric Approach*, UCL Press, London.

Ecologically Sustainable Development Working Groups (1991) *Final Reports*, AGPS, Canberra.

Economist, The (1994) 10 Sept., Vol. 332: 51–2.

—— (1995) 10 June, Vol. 335: 46.

Edelman, M. (1964) *The Symbolic Uses of Politics*, University of Illinois Press, Chicago.

Ekins, P. (1992) *A New World Order: Grassroots Movements for Global Change*, Routledge, London.

Fagin, A. (1994) 'Environmentalism and Transition in the Czech Republic', *Environmental Politics*, Fall, 3: 479–94.

Ferry, L. (1992) *The New Ecological Order*, The University of Chicago Press, Chicago and London.

Fischer, K. and Schot, J. (1993) *Environmental Strategies for Industry: International Perspectives on Research Needs and Policy Implications*, Island Press, Washington DC.

Foreman, D. (n.d.) 'A Spanner in the Woods', interviewed by B. Devall, *Simply Living*, 2, 12: 40–3.

Foucault, Michel (1984) *History of Sexuality, Volume One*, Peregrine Books, London.

—— (1991) 'Governmentality', chapter 4 in Graham Burchell, Colin Gordon and Peter Miller (eds) *The Foucault Effect: Studies in Governmentality*, Harvester Wheatsheaf, London.

Fox, W. (1990) *Toward a Transpersonal Ecology: Developing New Foundations for Environmentalism*, Shambala, Boston.

Gedicks, Al (1993) *The New Resource Wars: Native and Environmental Struggles Against Multinational Corporations*, South End Press, Boston.

Ghai, D. and Vivian, J. (1992) *Grassroots and Environmental Action: People's Participation in Sustainable Development*, Routledge, London and New York.

Global Environmental Facility Black Sea Environmental Programme (1994) *International Black Sea NGO Forum Meeting*, Report, 7–10 November.

—— (1995) *International Black Sea NGO Forum* Conclusions and Recommendations, 16–18 October.

—— Programme Coordination Unit (1995) *Commercial Fisheries in the Black Sea: Three Decades of Decline*, Leaflet.

—— (1995) *Black Sea NGO Directory*, Directory of Environmental NGOs in Bulgaria, Georgia, Romania, Russia, Turkey and Ukraine.

—— (1995) *Saving the Black Sea*, Official Newsletter, October, Issue 3.

Goldberg, K. (1994) 'Green relief for forest defenders', *The Progressive*, March, 58, 3: 13.

Goodwin, B. (1992) *Green Political Theory*, Polity Press, Cambridge.

Gore, A. (1992) *Earth in Balance: Ecology and the Human Spirit*, Houghton Mifflin, Boston.

Hansen, C. (1995) 'Sierra Club Management Shaming Muir's Memory', *Earth First!* Eostar edition: 26.

Hardin, G. (1968) 'The Tragedy of the Commons', *Science*, Vol. 162: 1,243–8.

Hartshorn, G.S. (1991) 'Key Environmental Issues for Developing Countries', *Journal of International Affairs*, Vol. 3: 393–402.

Hawken, P. (1993) 'A Declaration of Sustainability: 12 Steps Society Can Take to Save the Whole Enchilada', *UNTE Reader*, September/October: 54–61.

Hay, P. and Haward, M. (1988) 'Comparative Green Politics: Beyond the European Context?', *Political Studies*, 36: 433–48.

Hayes, G. (1994) 'In Splendid Isolation: The French Greens', *Environmental Politics*, 3: 169–73.

Hawken, P. (1993) *The Ecology of Commerce: A Declaration of Environmental Sustainability*, Harper Business, New York.

Helvarg, D. (1994) *The War Against the Greens: The "Wise Use" Movement, the New Right, and Anti-Environmental Violence*, Sierra Club Books, San Francisco.

Heschel, S. (1995) 'Feminists Gain at Population Conference', *Dissent*, Winter 1995: 15.

Hindess, Barry (1996) *Discourses of Power: From Hobbes to Foucault*, Basil Blackwell. Oxford.

Hirsch, P. and Lohman, L. (1989) 'Contemporary Politics of Environment in Thailand', *Asian Survey*, Vol. xxix, No. 4 (April): 439–51.

Holmes, R. and Kenny, M. (1995) 'The Electoral Breakthrough of the Irish Greens?', *Environmental Politics*, 3: 218–26.

Hornborg, A. (1994) 'Environmentalism, Ethnicity and Sacred Places: Reflection on Modernity, Discourse and Power', *The Canadian Review of Sociology and Anthropology*, 31, 008–4948: 245–67.

Homer-Dixon, T. (1994) 'Population and Conflict', distinguished lecture series on population and development, International Union for the Scientific Study of Population, Belgium.

Hunter, B., Siegfried, R., and Sunter, C. (1989) *South African Environments into the 21st Century*, Human & Rousseau and Tafelberg, Cape Town.

Hutchings, V. (1994) 'Green Gauge: Racism Plays a Key Role in Who Gets Toxic Waste Dumped on Them', *New Statesman and Society*, 11 March: 31.

—— (1994) 'Support Your Local Village Green', *New Statesman and Society*, March 11, 7, 293: 20–2.

Hyndman, D. (1991) 'Zipping Down the Fly on the OK Tedi Project', chapter 12 in John Connell and Richard Howitt (eds) *Mining and Indigenous Peoples in Australasia*, Sydney University Press, Sydney.

Imura, H. (1994) 'Japan's Environmental Balancing Act', *Asian Survey*, xxxiv, 4: 355–68.

Inglehart, R. (1977) *The Silent Revolution*, Princeton University Press, NJ.

—— (1990) *Culture Shift in Advanced Industrial Society*, Princeton University Press, NJ.

International Union for Conservation of Nature and Natural Resources [IUCN], United Nations Environment Programme [UNEP] and the World Wildlife Fund [WWF] (1980) *World Conservation Strategy: Living Resource Conservation for Sustainable Development*, Morges, Switzerland.

Jaensch, D. (1983) *The Australian Party System*, George Allen & Unwin, Sydney.

Jahn, D. (1994) 'Unifying the Greens in a United Germany', *Environmental Politics*, Vol. 3, No. 2 (Summer).

Jesinghausen, M. (1995) 'General Election to the German Bundestag on 16 October 1994: Green Pragmatists in Conservative Embrace or a New Era for Green Party Democracy', *Environmental Politics*, Spring, Vol. 4, No. 1: 108–14.

Johnson, S. (1995) *The Politics of Population: The International Conference on Population and Development Cairo 1994*, Earthscan Publications, London.

Joppke, C. and Markovits, A. (1994) 'Green Politics in the New Germany: The Future of an Anti-Party', *Dissent*, Spring: 235–40.

Jordan, A. (1994) 'Reviewing Rio: A Plummet from the Summit?', *Environmental Politics*, Spring: 159–63.

Kagin, M. (1995) 'Alternative Politics: Is a Third Party the Way Out', *Dissent*, Winter: 22–6.

Kuhn, T.S. (1969) *The Structure of Scientific Revolutions*, University of Chicago Press, Chicago.

Kramer, D. (1994) 'The Graying of the German Greens: Environmental Politics and the Crisis of Consensus', *Dissent*, Spring: 231–4.

Larana, E., Johnston, H. and Gusfield, J.R. (eds) (1992) *New Social Movements: From Ideology to Identity*, Temple University Press, Philadelphia.

Leftwich, A. (2983) *Redefining Politics: People, Resources and Power*, Methuen, London.

Lewis, M.W. (1992) *Green Delusions: An Environmentalist Critique of Radical Environmentalism*, Duke University Press, Durham, NC.

—— (1996) 'Radical Environmental Philosophy and the Assault on Reason', in P. Gross, N. Levitt and M.W. Lewis (eds) *The Flight from Science and Reason*, Annals of the New York Academy of Science, 775.

Lindblom, Charles (1977) *Politics and Markets*, Basic Books, New York.

List, P.C. (1993) *Radical Environmentalism: Philosophy and Tactics*, Wadsworth Publishing Company, Belmont, Calif.

Lowe, E. (1996) 'Industrial Ecology: A Context for Design and Decision', in J. Fiskel (ed.) *Design for Environment*, McGraw-Hill, New York.

Lowe, P. and Goyder, J. (1983) *Environmental Groups in Politics*, Allen & Unwin, London and Boston.

Lukes, Steven (1974) *Power: A Radical View*, Macmillan, London.

MacAndrews, C. (1994) 'Politics of the Environment in Indonesia', *Asian Survey*, XXXIV, 4: 369–80.

McEachern, Doug (1980) *A Class Against Itself: Power and the Nationalisation of the British Steel Industry*, Cambridge University Press, Cambridge, ch. 2.

—— (1991), *Business Mates: The Power and Politics of the Hawke Era*, Prentice Hall, Sydney.

—— (1993), 'Environmental Policy in Australia 1981–1991: A Form of Corporatism?', *Australian Journal of Public Administration*, Vol. 52, No. 2: 173–86.

—— (1995) 'Mining Meaning from the Rhetoric of Nature – Australian Mining Companies and Their Attitudes to the Environment at Home and Abroad', *Policy Organisation and Society*, No. 10 (Winter): 48–69.

—— (1997) 'Foucault, Governmentality and the "New" South Africa' in P. Ahluwalia and P. Nursey-Bray (eds) *Postcolonialism: Culture and Identity in Africa*, Nova Publishers, New York.

Maddox, John (1972) *The Doomsday Syndrome*, Macmillan, London.

Malthus, T.H. (1826) *An Essay on the Principles of Population* (sixth edition), Murray, London.

Martin, B. (1984) 'Environmentalism and Electoralism', *The Ecologist*, 14, 3: 110–18.

Maslow, A. (1954) *Motivation and Personality*, Harper and Row, New York.

Marx, Karl and Engels, Friedrich (1848) *Kommunistische Manifest [The Communist Manifesto]* with an introduction by A.J.P. Taylor [translated from the German by Samuel Moore], Penguin Books, Harmondsworth [1972].

Marx, K. (1967) *Capital*, 3 vols, International Publishers, New York.

Matthews, F. (1988) 'Deep Ecology: Where all Things are Connected', *Habitat*, October: 9–12.

Mayer, H. (ed.) (1969) *Australian Politics: A Second Reader*, Cheshire, Melbourne.

Miller, Peter and Rose, Nikolas (1993) 'Governing Economic Life', in Mike Gane and Terry Johnson (eds) *Foucault's New Domains*, Routledge, London, 75–105.

Meadows, Donella H. *et al.* (1972) *The Limits to Growth: A Report for the Club of Rome's Project on the Predicament of Mankind*, Potomac Associates, London.

Mellor, M. (1994) Book review of: *The Politics of the Environment* by Stephen C. Young in *Environmental Politics*, Fall, 3: 536–7.

Merchant, C. (1992) *Radical Ecology: The Search for a Livable World*, Routledge, New York and London.

Merchant, C. (ed.) (1994) *Ecology: Key Concepts in Critical Theory*, Humanities Press, New Jersey.

Mies, M. and Shiva, V. (1993) *Ecofeminism*, Fernwood Publications, Nova Scotia.

Naess, A. and Rothenberg, D. (1989) *Ecology, Community and Lifestyle*, Cambridge University Press, Cambridge.

National Wildlife Federation, 'NAFTA and the Environmental Side Agreements: Statement of Dr Jay D. Hair, President and CEO National Wildlife Federation, Washington DC', press release (14 September 1993).

North, Richard (1995) *Living on a Modern Planet: A Manifesto for Progress*, Manchester University Press, Manchester.

Oberschall, A. (1993) *Social Movements: Ideologies, Interests and Identities*, Transaction Publishers, New Brunswick and London.

O'Connor, K. and Sabato, L.J. (1995) *American Government: Roots and Reform* (second edition), Allyn and Bacon, Mass.

Paehlke, R. and Torgerson, Douglas (eds) (1990) *Managing Leviathan: Environmental Politics and the Administrative State*, Belhaven Press, London.

Paehlke, R. and Vaillancourt Rosenau, P. (1993) 'Environment/Equity: Tensions in North American Politics', *Policy Studies Journal*, 21, 4: 672–86.

Pakulski, J. (1991) *Social Movements: The Politics of Moral Protest*, Longman Cheshire, Melbourne.

Papadakis, E. (1993) *Politics and the Environment: the Australian Experience*, Allen & Unwin, Sydney.

Parsons, Talcott (1957) 'The Distribution of Power in American Society', *World Politics*, Vol. 10, 123–43.

Pearce, David (ed.) (1991) *Blueprint 2: Greening the World Economy*, Earthscan Publications, London.

Pepper, D. (1984) *The Roots of Modern Environmentalism*, Croom Helm, London.

—— (1993) *Eco-Socialism: From Deep Ecology to Social Justice*, Routledge, London and New York.

Piven, F.F. and Cloward, R.A. (1979) *Poor People's Movements: Why They Succeed, How They Fail*, Vintage Books, New York.

Plamenatz, J. (1973) *Democracy and Illusion*, Longman, London.

Plumwood, V. (1988) 'Women, Humanity and Nature', *Radical Philosophy*, 48: 6–24.

Poguntke, T. (1992) 'Unconventional Participation in Party Politics: The Experience of the German Greens', *Political Studies*, XL: 239–54.

Polsby, Nelson W. (1963) *Community Power and Political Theory*, Yale University Press, New Haven, Conn., and London.

Porritt, J. (1984) *Seeing Green: The Politics of Ecology Explained*, Basil Blackwell, New York and Oxford.

Porter, G. and Welsh Brown, J. (1991) *Global Environmental Politics: Dilemmas in World Politics*, Westview Press, Boulder, Colo.

Powers, M. (1993) 'Now What? The Environment', *Human Ecology Forum*, 21, 1: 20–3.

Prendiville, B. (1994) *Environmental Politics in France*, Westview Press, Boulder, Colo., San Francisco and Oxford.

President's Council on Sustainable Development (1996) *Final Report*, [http://www.whitehouse.gov/WH/EOP/pcsd/Council_report], March.

Preston-Whyte, R.A. (1995) 'The Politics of Ecology: Dredge-mining in South Africa', *Environmental Conservation*, Vol. 22, No. 2: 151–6.

Princen, T. and Finger, M. (1994) *Environmental NGOs in World Politics: Linking the Local to the Global*, Routledge, London and New York.

Richardson, D. and Rootes, C. (1995) *The Green Challenge: The Development of Green Parties in Europe*, Routledge, London and New York.

Richardson, G. (1994) *Whatever It Takes*, Bantam Books, Sydney.

Rohrschneider, R. (1991) 'Public Opinion Toward Environmental Groups in Western Europe: One Movement or Two?', *Social Science Quarterly*, 72, 2: 251–66.

—— (1993) 'Environmental Belief Systems in Western Europe: A Hierarchical Model of Constraint', *Comparative Political Studies*, 26, 1: 3–29.

Rosen, R. (1994) 'Who Gets Polluted: The Movement for Environmental Justice', *Dissent*, Spring: 223–30.

Rosen, S.J. and Nolan, T. (1994) 'Seeking Environmental Justice for Minorities and Poor People', *Trial*, December: 50–5.

Rosenbaum, Walter A. (1991) *Environmental Politics and Policy* (second edition), Congressional Quarterly Inc., Washington DC.

Rudig, W. and Franklin, M. (1992) 'Green Prospects: The Future of Green Parties in Britain, France and Germany', in W. Rudig (ed.) *Green Politics Two*, Edinburgh University Press, Edinburgh: 37–58.

Salleh, A. (1992) 'The Ecofeminism/Deep Ecology Debate: A Reply to Patriarchal Reason', *Environmental Ethics*, Fall edition: 195–216.

Sale, K. (1993) *The Green Revolution: The American Environment Movement 1962–92*, Hill and Wang, New York.

Sandbrook, R. (1993) 'Live and Learn', *New Statesman and Society*, January: 29–30.

Scarce, R. (1990) *Eco-Warriors: Understanding the Radical Environmental Movement*, The Noble Press, Chicago.

Schrader, R. (1992) 'The Fiasco at Rio', *Dissent*, Fall 1992: 431.

Schrecker, Ted (1990) 'Resisting Environmental Regulation: The Cryptic Pattern of Business–Government Relations', in Robert Paehlke and Douglas Torgerson (eds) *Managing Leviathan: Environmental Politics and the Administrative State*, Belhaven Press, London (1990), 165–99.

Schreiber, H. (1995) 'The Threat for Environmental Destruction in Eastern Europe', *Journal of International Affairs*, 361–91.

Seliger, M. (1976) *Ideology and Politics*, Allen & Unwin, London.

Sessions, G. and Naess, A. (1983) 'The Basic Principles of Deep Ecology', *Earth First*.

Shiva, V. (1994) 'Development, Ecology and Women', in C. Merchant (ed.) *Ecology: Key Concepts in Critical Theory*, Humanities Press, New Jersey.

Simon, Julian and Kahn, Herman (1984), *The Resourceful Earth*, Basil Blackwell, Oxford.

Skinner, J. (1995) 'Green Party Revival', *The Progressive*, May, 59, 5: 14.

Snow, D. (ed.) (1992a) *Inside the Environmental Movement: Meeting the Leadership Challenge*, Island Press, Washington DC and Covelo, Calif.

—— (ed.) (1992b) *Voices from the Environmental Movement: Perspectives for a New Era*, Island Press, Washington DC and Covelo, Calif.

Szabo, M. (1994) 'Greens, Cabbies and Anti-Communists: Collective Action During the Regime Transition in Hungary', in E. Larana *et al.* (eds) *New Social Movements: From Ideology to Identity*, Temple University Press, Philadelphia.

Taylor, B. (1991) 'The Religion and Politics of Earth First!', *The Ecologist*, Vol. 21, No. 6, Nov./Dec.: 258–66.

The 3rd Citizen's Conference on Dixon and Other Synthetic Hormone Disruptors (1996) *Time for Action*, Baton Rouge, La.

Thomas, Ian (1996) *Environmental Impact Assessment in Australia: Theory and Practice*, The Federation Press, Sydney.

Thompson, H. (1993) 'Malaysian Forestry Policy in Borneo', *Journal of Contemporary Asia*, 23, 4: 503–14.

United Nations Conference on Environment and Development (1992), *Agenda 21*, Geneva (on line at http://www.erin.gov.au/portfolio/esd/nesd/Agenda21/html).

Ward, Barbara and Dubos, Rene (1972) *Only On Earth*, W.W. Norton, New York.

Weatherly, C. (1993) 'The Ecological Imperative and Political Success: A Paper for Ecopolitics VII', Abstract for Ecopolitics VII Conference, 2 July 1993.

Weber, Max (1978) *Economy and Society: An Outline of Interpretive Sociology* (edited by G. Roth and C. Wittich), University of California Press, Berkeley.

Women's Environment and Development Organization (WEDO) (1995) 'Transnational Corporations at the UN: Using or Abusing their Access?', newsletter, 2: 2–4.

World Commission on Environment and Development [WCED, chaired by Gro Harlem Brundtland] and the Commission for the Future (1990) *Our Common Future* (Australian edition), Oxford University Press, Melbourne.

Young, J. (1990) *Post Environmentalism*, Belhaven Press, London.

Index